Beyond Strength: Addressing the Durability Challenge in Concrete Structures

Namita

Contents

1. Introduction

1.1 Background

The availability, ease of use and effectiveness of Portland cement has made concrete the primary choice of building material, especially in developing countries [4]. In current construction practice, it is assumed that the mechanical properties of different types of concrete have been considerably investigated and are well understood. However, the durability of concrete structures has not been sufficiently addressed; having been assumed to be an inherent property of the concrete. Concerns about durability arose when modern structures were witnessed to experience significant deterioration during their service lives [5].

An effort to understand and mitigate the mechanisms by which deterioration takes place has highlighted the knowledge gap in designing durable concrete structures. Deterioration typically occurs as a result of the penetration of deleterious substances into the concrete [6]. Although there are various forms of concrete degradation, the most significant is the corrosion of steel in reinforced concrete [6, 7, 8].

Reinforcement corrosion is an electrochemical process that occurs either as carbon dioxide-induced or chloride-induced corrosion depending on the aggressive agent [5]. Research shows that the latter is the more aggressive of the two since it occurs at a higher rate, resulting in failure within a shorter time after initiation [9].

The most severe exposure condition in Cape Town in the Western Cape province of South Africa is considered to be the marine splash and spray zone [9]. Reinforced concrete structures in the marine environment are most aggressively affected by chloride-induced corrosion [9]. Chlorides in this environment are present as salts dissolved in sea water and mist. For chloride-induced corrosion to occur, it is necessary for chloride ions to penetrate through the cover concrete to reach the reinforcing steel surface. This transportation of fluids through the pores and cracks depends on the penetrability of the concrete in question [5]. The presence of oxygen and moisture at the steel surface is additionally necessary for the initiation and propagation of reinforcement corrosion. Cyclic wetting and drying in the marine splash and spray zone provides suitable conditions for sustaining chloride-induced reinforcement corrosion.

The products of corrosion occupy a volume up to six times that of the intact steel. They are accommodated in the interfacial transition zone between the concrete and the steel [10]. The volumetric increase results in loss of bond between the concrete and the steel and creates expansive tensile stresses within the concrete [11]. This eventually leads to cover cracking, concrete delamination and spalling. These phenomena compromise general safety and the repair and maintenance of structures affected by reinforcement corrosion is significantly costly [9].

In order to address the problem of reinforced corrosion in concrete, it is necessary to combine experimental studies, analytical solutions and numerical methods to predict the transportation of deleterious species through the concrete and the accompanying stresses. The results may then be used to more accurately predict the course of material degradation so that the available mitigation measures such as cathodic protection may be used more effectively [12].

Service life models for reinforced concrete are limited by the various complications presented by the corrosion process. These include the variation in the composition and density of the corrosion products, the non-uniform distribution around the steel reinforcing of said products and penetration of the corrosion products into the pores of the concrete. Significant inconsistencies are often observed between the modelled and the measured time-to-cracking [3, 13]. Val et al. [2] discerned these differences to stem from not adequately accounting for the penetration of rust into the concrete pore spaces next to the reinforcing steel. Due to the penetration of rust into the concrete pores, a greater steel mass loss is required to induce expansive stresses at the steel-concrete interface. This translates to a longer corrosion propagation period before cracking of the concrete cover occurs [10].

Existing models that have attempted to include the penetration of iron III chloride typically assume a uniform corrosion product with uniform distribution around the steel and uniform penetration into the concrete pores at the steel surface. In some of these models, the penetration of corrosion products into the concrete pores is addressed separately from the formation of corrosion induced cracks. By considering these two phenomena separately, the influence of the penetration of corrosion products into the concrete pores on the cracking of concrete due to reinforcement corrosion cannot be determined. The models reviewed did not directly quantify the volume of corrosion products penetrating into the concrete pores or the depth of penetration. These parameters were rather calculated from other variables in the model such as corrosion current density and capillary porosity.

Concrete is a composite, porous material with a solid skeleton made up of the hydrated cement paste and various sized pores [4]. It typically contains micro cracks as well, resulting from shrinkage and differential settlement stresses [14]. The pores and cracks provide channels through which fluids such as air and water may be transported. The movement of fluids through the pore structure is governed by different mechanisms such as diffusion and capillary suction depending on the conditions to which the concrete is subjected to [5].

Multiphase modelling is used to model fluid flow in a solid body containing one or more fluids [15, 16]. It was originally developed for application to soil mechanics as elaborated in de Boer [16] but has been adapted for petroleum and chemical engineering, materials science and biomechanics [17] due to advancements in mechanics of porous media.

The Theory of Porous Media (TPM) is a framework for multiphase modelling of porous bodies [16, 18]. It is based on classical mixture theory restricted by volume fractions [16]. A biphasic TPM growth model for application to Rheumatic Heart Disease was recently developed at the University of Cape Town and implemented in the in-house software *SESKA* by Hopkins [19]. The results of the study demonstrated the efficacy of TPM in modelling complex transport phenomena in a porous body. Considering the developments in multiphase modelling of cementitious materials [20, 21, 22], it is envisaged that TPM may be suitably used to model the processes involved in chloride-induced corrosion of reinforced concrete structures.

1.2   Aim of the study

Attempts have previously been made to include the penetration of corrosion products in time-to-cracking service life models for reinforced concrete subjected to corrosion [1, 2, 3] in an endeavour to produce more accurate results. However no attempts have yet been made to model the mechanism by which the penetration occurs. This may be attributed to a lack of experimental studies on the subject and therefore a deficiency in the understanding of the precise mechanism. At present, it is assumed that the corrosion products penetrate through the concrete capillary pores by diffusing through the pore solution and precipitating in the capillary pores close to the steel surface [10].

Multiphase porous media theory has recently been applied to deterioration processes in concrete such as calcium leaching and alkali silica reaction [20, 23, 24]. These studies demonstrate the efficacy of multiphase modelling for the intricate coupled phenomena occurring during concrete deterioration. However, the theory has yet to be applied to corrosion in reinforced concrete.

Recognising the possible effectiveness of multiphase modelling for chloride-induced corrosion in reinforced concrete, the aim of this study was to develop a numerical model based on the Theory of Porous Media that may be used to simulate the penetration of corrosion products through the concrete pore structure. The kinetics of corrosion were not explicitly addressed in the model as the focus was deemed to be on the transport mechanisms involved in the corrosion process.

This is the first step in the development of a more accurate service life model based on the Theory of Porous Media for chloride-induced corrosion of reinforced concrete.

1.3 Objectives of the study

The objective of this study was to derive the theoretical framework for a triphasic numerical model using the Theory of Porous Media scheme to calculate the deformations resulting from the transport of a fluid and a solute through the porous medium. This would be adapted to the penetration of corrosion products through the reinforced concrete capillary pores by adjusting the material properties in the model to those of concrete.

The main objective was broken down into smaller objectives deemed achievable in the limited time-frame of this study and these are;

- To investigate the relationship between transport mechanisms in concrete, corrosion rate, rust development and transport.
- To derive a triphasic framework based on the Theory of Porous Media for fluid transport in a porous solid.
- To add a miscible component (called the solute ion) into the model that is transported through the wetting fluid (called the solvent) by the process of diffusion.

1.4   Scope and limitations

A theoretical framework for a fluid and ion transport model based on TPM is developed for a partially saturated solid. The ion is included as a soluble constituent in the wetting fluid phase. The model couples the transport process with the poro-elastic material response of the solid body. Mass exchange between the constituents of the porous medium is not incorporated and therefore the model does not take into account the precipitation of the soluble ion in the solid pores. It is assumed there is a continuous external supply of the solute ions.

The sorption-desorption isotherm and associated parameters employed in this work result in a mathematical ambiguity for negative capillary pressures. Negative capillary pressures may occur if the wetting fluid is imbibed into a completely dry porous solid. Hence the application of the model presented is limited to drainage and imbibition of a partially saturated solid body.

The model presented is generic and needs to be adapted for reinforced concrete by calibration and validation using material properties and experimental data for concrete. Calibration and validation in *SESKA* could not be executed in the limited time frame of this study, however these should be done in order to determine the effectiveness of the model for simulating the penetration of corrosion products into the concrete pores.

1.5   Layout of the document

Chapter 1 of the document provides a brief introduction and background information to the research study conducted. Chapter 2 is a detailed literature review concerned with reinforcement corrosion, the penetration of corrosion products into the concrete pore spaces and the modelling of these two phenomena. Chapter 3 details the generic Theory of Porous Media framework which is adapted in Chapter 4 for a three phase system with a solute in the liquid phase. The first attempts at numerical implementation together with the results of the simulations are presented in Chapter 5. Finally, Chapter 6 gives the conclusion to this document and details the work still to be done.

2. Literature review

This chapter presents the literature review conducted for this study. It gives an account of the relationship between service life and durability and how these are affected by chloride-induced corrosion. A further investigation was conducted on how reinforcement corrosion occurs and what transport mechanisms in concrete are involved in the process. In exploring the effect of chloride-induced corrosion on the service life of reinforced concrete structures, it was necessary to identify and understand the material properties of corrosion products. The literature study then focused on how the corrosion products lead to loss of service life, how this process may be modelled and the shortcomings in existing models of this phenomenon.

2.1 Overview

This section gives a brief introduction to the concepts of service life and durability, mechanisms of deterioration of reinforced concrete and corrosion of reinforcing steel in concrete.

2.1.1 Service life and durability of concrete

A qualitative definition of service life is given by Helland [25] as the duration over which a structure or any of its components is used and maintained as intended, without the requirement for major repairs. Ballim et al. [5] define durability of a concrete structural component as its capacity to endure the exposure environment over its design life. A material designed for a specific environment may not be durable in an alternative environment and therefore durability is a correlation between the material and its service environment [5]. Service life design is employed for realising structural concrete durability and is traditionally included in design standards.

It is estimated that up to 40% of construction budgets in developed countries are spent on the repair, rehabilitation and maintenance of concrete structures [4, 26]. This comes at a significant

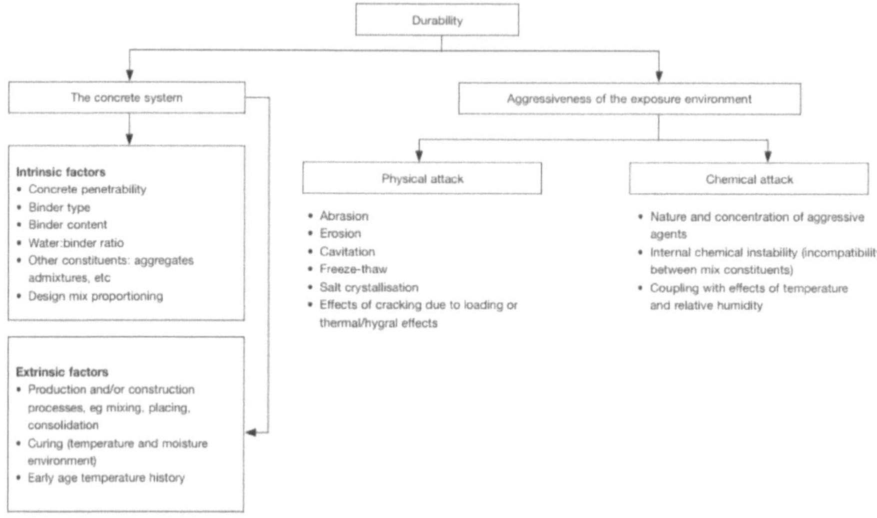

FIGURE 2.1: Factors affecting durability [5].

socio-economic cost that has led to cognizance of the importance of durability design and life-cycle versus initial costs of new establishments. It has also been discerned that use of durable materials results in conservation of natural resources and is therefore advantageous for the environment [4].

Ballim et al. [5] classify factors affecting durability into factors that determine how the concrete resists degradation and environmental factors that determine the severity of the conditions the concrete structure is exposed to, as shown in Figure 2.1. Factors affecting how concrete resists degradation are divided into intrinsic and extrinsic factors. Intrinsic factors include binder type and water/binder ratio. Extrinsic factors include ambient temperature, construction quality and curing. Environmental factors affecting concrete degradation include the presence of aggressive agents such as chlorides and carbon dioxide [5].

2.1.2 Mechanisms of deterioration of reinforced concrete

Deterioration mechanisms of reinforced concrete are intricate phenomena spanning multiple length scales and various fields of study [4]. These mechanisms generally involve the transportation of fluids through the concrete pore structure [5]. The probability of occurrence of a specific type of deterioration and the rate at which it progresses depends on the severity of the exposure conditions, the resistance of the concrete to these conditions and the fluid-transport attributes of the concrete [5].

Almost all deterioration mechanisms of concrete require the action of water, whether present in the exposure environment or within the concrete pores. Water enables dissolved salts to partake in ion-exchange and addition reactions with concrete [4, 5].

Mehta and Monteiro [4] classify deterioration mechanisms as either physical or chemical processes. The physical processes are further subdivided into "surface wear" and "cracking". Surface wear may occur as a result of abrasion, erosion and cavitation.

Reinforced concrete members may crack in high load intensity areas due to stresses from applied load [5]. These micro-cracks are admissible in the normal service life of the structure and typically do not cause concern. Cracking in concrete may also occur as a result of freezing of pore water, fire and shrinkage-induced stresses as water is evaporated from the concrete surface. The presence of cracks in concrete increases the rate at which deleterious substances penetrate into the concrete [5].

Chemical deterioration processes may be generally classified into ion exchange, ion removal or ion addition reactions [5]. Ion exchange reactions involve the action of acids on the alkaline cement paste matrix resulting in consumption of the hydrated cement paste. Ion removal entails the dissolution and leaching of the hydrated cement paste. The reaction of the hydrated cement paste with deleterious substances to form expansive reaction products is an ion addition process. The reaction products induce expansive stresses within the concrete and cause cracking of the concrete [5].

The aforementioned deterioration mechanisms typically do not occur in isolation but tend to manifest with and intensify other mechanisms.

2.1.3   Corrosion of reinforcing steel

Corrosion of reinforcing steel in concrete is a unique form of deterioration [5]. It is a major limiting factor in the service life and durability of reinforced concrete structures. It occurs through the mechanism of oxidation of iron from carbon steel to produce expansive oxides of the consumed metal. The accumulation of these products at the steel-concrete interface exerts pressure on the cover concrete, which eventually leads to debonding between the concrete and steel, concrete cover-cracking and spalling.

Reinforcement corrosion in concrete occurs as either carbonation-induced or as chloride-induced corrosion [2, 4, 5]. Chloride-induced corrosion is the foremost concern as it occurs at a much higher rate and results in pitting of the reinforcing steel [7, 9, 11]. Structures in the marine environment and those exposed to de-icing salts are the most susceptible to reinforcement corrosion [9].

Carbonation-induced corrosion occurs when carbon dioxide is allowed to diffuse into concrete from the atmosphere and reacts with products of the hydration of cement to form calcium carbonate [27]. This results in the reduction of pH of the concrete matrix and hence the breakdown of the passive oxide layer on the steel [28]. The loss of passivation of the reinforcing steel initiates corrosion. Relative to chloride-induced corrosion, this type of corrosion has a lower reaction rate. Carbonation will not occur if the concrete pores are completely saturated or completely dry. It typically occurs at a relative moisture content between 50% and 70% [5].

Chloride-induced corrosion requires contact between the reinforcing steel and chloride ions. The ingress of chloride ions into reinforced concrete depends on the penetrability of the concrete [5]. Concrete binds chloride ions to form a complex called *Friedel's salt* ($[Ca_2Al(OH)_6]Cl.2H_2O$) [29]. Free chlorides are chloride ions not bound in the concrete. A critical concentration of free chloride ions at the steel level results in the loss of the passive oxide layer on the steel surface and corrosion is initiated thereafter [5].

The focus of this work is on chloride-induced corrosion and therefore a detailed discussion on carbonation-induced corrosion will not be entered into. Chloride-induced corrosion is elaborated in Section 2.5.

2.2 Concrete micro-structure

Advancements in materials science have led to appreciation of the influence of the internal micro-structural properties on the macro-structure of concrete [30, 31]. However, the high complexity of this micro-structure makes it difficult to develop efficient and realistic models for the prediction of concrete material behaviour. Understanding of the micro-structural properties of the individual constituents of concrete and their interactions is important when attempting to develop structural and material behavioural models [4].

The microstructure of concrete may be characterised by the hydrated cement paste, the pore structure and the ITZ (Interfacial Transition Zone) [4, 32]. At the micro-structural level, the hydrated cement paste and ITZ are observed to be inhomogeneous both inherently and in relation to each other. Each of these constituents generally contain varying quantities of solid phases, pore spaces and micro-cracks. The micro-structural constituents of concrete vary with time, humidity and temperature and therefore the micro-structure is not considered an intrinsic characteristic of the concrete [4].

2.2.1 Hydrated cement paste (hcp)

When cement comes into contact with water, a chemical reaction occurs that results in the development of the hydrated paste . The different components that make up the paste micro-structure are variable in distribution, size and morphology. These variations significantly affect the mechanical and physical properties of the concrete [4].

Particles of unhydrated cement may be found in almost all hcps as shown in Figure 2.2. Hydrated cement pastes commonly consist of Calcium-Silicate-Hydrate (CSH) gel, calcium hydroxide, a hydrous calcium aluminium sulfate called ettringite and monosulfate interlaced with pore spaces. The CSH gel is a product of Alite (Ca_3SiO_5) and Belite (Ca_2SiO_4) hydration in cement and varies within the hcp. As the concrete ages, hydration continues resulting in new hydration products filling up the pore spaces and making the hcp more dense [33].

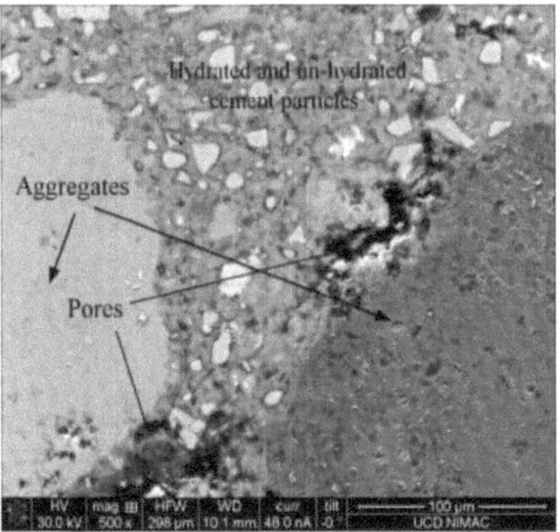

FIGURE 2.2: A Scanning Electron Microscopy (SEM) image of the concrete microstructure [34].

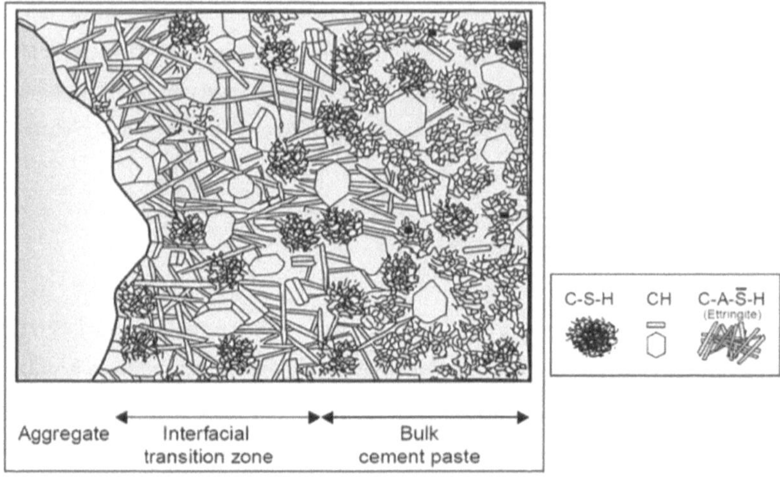

FIGURE 2.3: A schematic representation of the ITZ [4].

2.2.2 Interfacial Transition Zone (ITZ)

Concrete contains both fine and coarse aggregates. The bulk density and strength of the aggregate provide the concrete with dimensional stability and affect its unit weight and elastic modulus [5]. Because these are primarily physical characteristics, it is deduced that the chemical characteristics of the aggregate play a lesser role in the concrete except in the case of alkali-aggregate reactions. The shape and size of the coarse aggregate indirectly affect the strength and permeability of concrete. The higher the proportion of elongated and flat aggregate, the greater will be the tendency for water to filtrate through the ITZ [4].

The micro-structure of concrete is known to vary in zones close to large aggregate from the bulk paste as illustrated in Figure 2.3. The ITZ is formed as a result of the composite nature of concrete between the aggregate and the hydrated cement paste. Hu [14] describes it as a packing discontinuity at the aggregate particle surface. This zone may be up to 50 μm thick and has been observed to have higher porosity and penetrability than the matrix. The porosity may be up to 30% in the region within 2 μm of the aggregate. This provides favourable paths for fluid transport and significantly influences the mechanical behaviour of the concrete [5, 33].

Winslow and Cohen [35] found that when the ITZ reached a size of $20\mu m$, percolation was initiated as shown in Figure 2.4. This is because at a sufficient volume of aggregate within the concrete, the

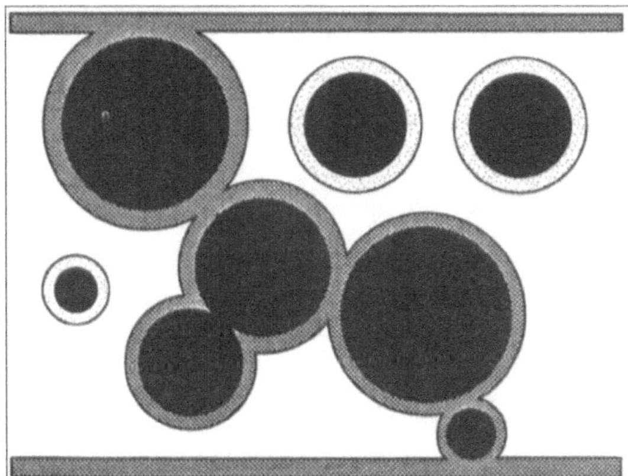

FIGURE 2.4: Percolation in concrete due to increasing size of ITZ [35].

ITZs begin to overlap creating an unhindered path for fluid transport. Percolation describes the connectivity of the pore spaces and factors into transport mechanisms within the concrete.

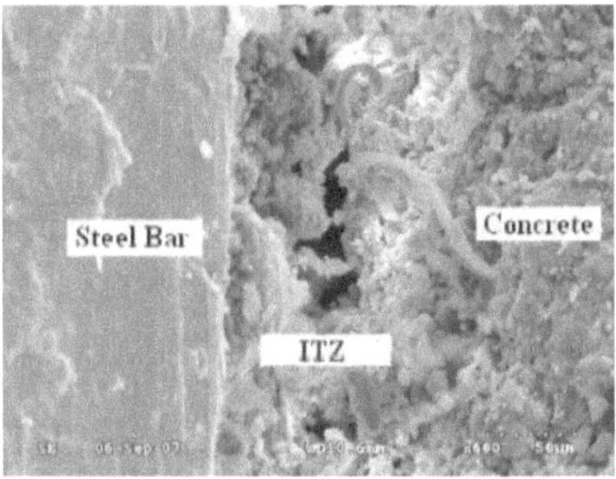

FIGURE 2.5: An SEM image of the ITZ between steel and concrete [36].

In reinforced concrete, an ITZ exists between the steel and the concrete matrix as shown in Figure 2.5 [3, 36]. It has been described as having similar characteristics to the ITZ between aggregate and the hcp [36]. The steel-concrete ITZ consists of capillary pores, exhibits significantly higher porosity than the cement matrix and offers room for the deposition of corrosion products during reinforcement corrosion [3]. This deposition phenomenon was observed (by comparing Figures 2.5 and 2.6), as the densification of the ITZ during accelerated corrosion experiments in which the ITZ was examined under an Scanning Electron Microscope (SEM) by Yuan and Ji [36].

FIGURE 2.6: An SEM image of the corrosion product layer between the steel and concrete [36].

The existence of an ITZ is more evident in younger concretes. This may be attributed to the absence of cement particles in the regions immediately surrounding the aggregate in fresh concrete. It results in a higher percentage of water-filled spaces close to the aggregates in younger as opposed to older concrete. These spaces are later filled by calcium hydroxide or CSH upon continued hydration of the cement paste [33].

2.2.3 Porosity of concrete

Hu [14] defines porosity as the ratio of pore volume to the initial paste volume. Four types of pores in concrete have been identified namely: gel pores, capillary pores, macro pores resulting from entrained air and macro pores resulting from insufficient compaction [14, 32]. The variation in pore size from nanometres to millimetres is shown in Figure 2.7.

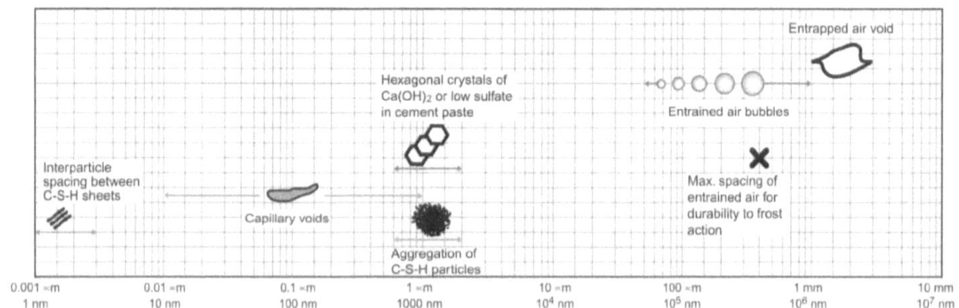

FIGURE 2.7: Types of pores present in the hydrated cement paste [4].

Concrete porosity depends on various factors, the most important of which are; the degree of hydration of the concrete, particle size distribution of the aggregate and the water-binder ratio.

The packing discontinuity of the ITZ as described in Section 2.2.2 also influences the porosity of concrete [14].

Porosity significantly affects permeability of concrete and hence its strength and durability. Although there is no direct correlation between porosity and permeability, it has previously been established that as the cement hydration reaction proceeds, both capillary porosity and permeability decrease [4].

2.3   Transport mechanisms in concrete

Durability of concrete has been ascertained to depend on the transport of fluids and dissolved ions in the pore structure. Fluid and ion movement may be due to capillary suction, flow under pressure or a concentration gradient. It is for this reason that degradation processes such as chemical attack, leaching, chloride ingress or carbonation are influenced by the transport mechanisms of concrete. These mechanisms include permeation, absorption, diffusion and migration [5].

The transport mechanisms mentioned in this section often cannot be treated singularly in a given instance. In reality, these mechanisms may occur simultaneously within a given time or section of the member being addressed. These mechanisms vary with the age of the concrete due to continuing hydration of the cement paste or due to deterioration of the concrete. The variations affect the porosity of the concrete and this should be taken into account when developing models of transport mechanisms.

2.3.1   Permeation

This is the flow of a fluid through the pores of fully saturated concrete due to an externally applied pressure. The capacity of a porous body to permeate fluids is termed *permeability*. Mehta and Monteiro [4] define permeability as "the rate of viscous flow of a fluid under pressure through the pore structure". The coefficient of permeability is a fluid transport characteristic and for steady state flow is typically determined from *Darcy's* Law according to

$$\mathbf{q} = -\frac{\kappa}{\mu}(\boldsymbol{\nabla} p - \rho \mathbf{g}) \tag{2.1}$$

where $\mathbf{q}$ is the flux , κ is the permeability, μ is the viscosity, ∇p is the pressure gradient, ρ is the density of the fluid and $\mathbf{g}$ is the body force [37]. In porous media theory, the flux $\mathbf{q}$ is analogous to the advective mass flux expressed as a seepage velocity for each of the fluids [16, 37, 38].

The permeability of the hcp varies significantly depending on the degree of hydration and on the size, structure and connectivity of the pores, which are governed by the porosity. A small reduction in the porosity typically corresponds to substantial divisions of the larger pores. This reduces the size and continuity of the pores, hindering flow in the cement paste [4].

Research has shown that the coefficient of permeability for concrete is appreciably higher than that of a comparable mortar. The increase in the coefficient of permeability is attributed to the inclusion of large aggregate and is directly proportional to the size of aggregate used [4]. This phenomenon may be explained by the higher permeability of the ITZ between the aggregate and the hcp as previously explained.

2.3.2 Diffusion

The motion of species through a partially or fully saturated material due to a concentration gradient is called diffusion. The rate of diffusion through concrete depends on temperature, moisture content of concrete, and the species being diffused. Diffusion in concrete is influenced by: chemical interactions with the cement hydration products; by the saturation conditions; by cracks and voids in the concrete; and by electrochemical effects due to rebar corrosion [5].

Diffusion characteristics are vital in modelling deterioration of reinforced concrete structures in marine environments. *Fick's* first law of diffusion is typically used to model the diffusion of gases through concrete pores and solute ions within the pore solution. It is given by

$$J = -D_{eff}\frac{dC}{dx} \tag{2.2}$$

where J is the mass transport rate ($\frac{g}{m^2 s}$), D_{eff} is the effective diffusion coefficient ($\frac{m^2}{s}$), $\frac{dC}{dx}$ is the concentration gradient ($\frac{\frac{g}{m^3}}{m}$), C is the concentration of diffusing species in the fluid ($\frac{g}{m^3}$) and x is the distance (m). The chemical reactions are accounted for by adding a chemical reaction reaction term to the generic *Fick's* Law for each species being diffused [39].

Fick's second law of diffusion used to model ionic diffusion is given by

$$\frac{\partial C}{\partial t} = D\frac{\partial^2 C}{\partial x^2} \tag{2.3}$$

where D is the diffusion coefficient $(\frac{m^2}{s})$, t is the time (s).

2.3.3 Capillary absorption

Absorption is defined as the movement of a fluid through the pore spaces of an unsaturated material due to capillary suction. It is influenced by the arrangement of pores and the degree of saturation of the material. The rate of movement of a wetting front due to absorption is called *Sorptivity* [5].

Pérez et al. [40] defines capillary sorption and desorption as the flow of fluids from saturated to partially saturated regions. It occurs as a result of adhesive forces between water molecules and the walls of the concrete pores [41]. The fluids may carry along dissolved species that are described by their concentrations. Capillary absorption may be considered the prevailing transport mechanism in the vicinity of an unsaturated concrete surface since diffusion occurs at a much slower rate [42].

In partially saturated concrete, single phase flow resulting from capillary absorption may be given by an extension of the *Darcy* equation expressed as

$$\mathbf{q} = -\kappa(\theta)\,\boldsymbol{\nabla}\Psi(\theta), \tag{2.4}$$

where $\mathbf{q}$ is the flow velocity, κ is the hydraulic conductivity, Ψ is the capillary potential and θ is the reduced water content defined as

$$\theta = \frac{\Theta - \Theta_i}{\Theta_s - \Theta_i}. \tag{2.5}$$

In Equation (2.5), Θ is the volume fraction of water in the medium, Θ_i is the initial liquid volume fraction and Θ_s is the saturated liquid volume fraction so that θ is initially zero and is one at maximum saturation [43].

2.3.4 Transport mechanisms in cracked concrete

Cracked concrete exhibits different properties to that of uncracked concrete. In cracked concrete, the transport mechanisms are dependent on a correlation between the crack network and the concrete matrix. Crack width, origin, frequency and degree of connectivity all play a vital role in transport mechanisms in cracked concrete [9]. Both micro and macro cracks influence these mechanisms differently. Data based on intact concrete cannot be applied directly to cracked concrete. In some cases of cracked concrete, the properties of the cracks in relation to the transport mechanisms may become more important than the properties of the concrete itself [5]. For simplicity, this work assumes the concrete to be intact so that the influence of cracks on fluid transport through the concrete pores may be disregarded.

2.4   Moisture state

The main transport mechanisms used in this work are described in Section 2.3. These processes are significantly influenced by the degree of saturation of concrete with water. It is therefore essential to adequately account for the varying moisture contents experienced by real structures in their service life.

2.4.1   Capillary pressure and vapour pressure

Capillary pressure is the pressure difference between two immiscible fluid phases resident in the same pore space. It is caused by inter-facial tensile forces between the two fluids which must be exceeded for either fluid to flow [44]. There exists a relationship between fluid saturation and capillary pressure and this is elaborated in Section 2.4.2 [37].

Vapour pressure is the equilibrium pressure applied by a vapour on its liquid or solid form at a specified temperature in a closed system [45]. It is a measure of the tendency for molecules to break away from their more condensed form and therefore is related to the rate of evaporation of the liquid and condensation of the gas.

Partial saturation of the porous medium is defined as the presence of more than one fluid phase within the same pore space [22]. The moisture state variable chosen for the partially saturated porous medium needs to show acceptable numerical performance upon computation [22].

The established correlation between pressures and stress makes both vapour pressure and capillary pressure well-suited for the stress-state analysis of partially saturated reinforce concrete. The use of vapour pressure as the moisture state variable has the advantage of being valid at all temperatures, however it loses significance as a physical variable in fully saturated conditions [20] . On the other hand, capillary pressure is only valid at temperatures below the critical point of water and above the solid saturation point [20, 37]. The solid saturation point is the degree of saturation below which the liquid exists only as adsorbed water and above which the liquid phase exists both as adsorbed and capillary water [37]. Capillary pressure was chosen for this work. It was shown by Gawin et al. [20] to be equal to a thermodynamic water potential so that it could satisfactorily be used as the moisture state variable.

2.4.2   Sorption-desorption isotherm

The sorption-desorption isotherm graphically expresses capillary sorption and desorption phenomena in concrete [37]. It defines the relationship between saturation with the liquid, herein denoted as s^L, and the capillary pressure denoted as p^C. The relationship expressed by the isotherm is used to reduce the number of unknowns in the equation system for solution of the numerical model in this study. A sorption-desorption isotherm is characteristic of the micro-structure of the porous medium which in the case of concrete, varies with continued hydration of the cement paste [22].

The liquid saturation is the ratio of the liquid phase volume to the total pore volume in a differential volume element of the material. It may be evaluated from the capillary pressure using the most common sorption-desorption isotherm known as the *van Genuchten* [46] model according to the equation

$$s^{\mathbf{L}} = [1 + (\alpha p^C)^j]^{-h} \tag{2.6}$$

with j and h coupled according to the equation

$$h = 1 - \frac{1}{j} \tag{2.7}$$

where α and j are material constants [38]. The saturation degree is then used to determine the relative permeability which will be discussed in Section 2.4.3.

Typical capillary pressure behaviour in porous media is shown in Figure 2.8 and is categorized into drainage and imbibition. Drainage refers to a decrease in liquid saturation and imbibition refers to an increase in liquid saturation.

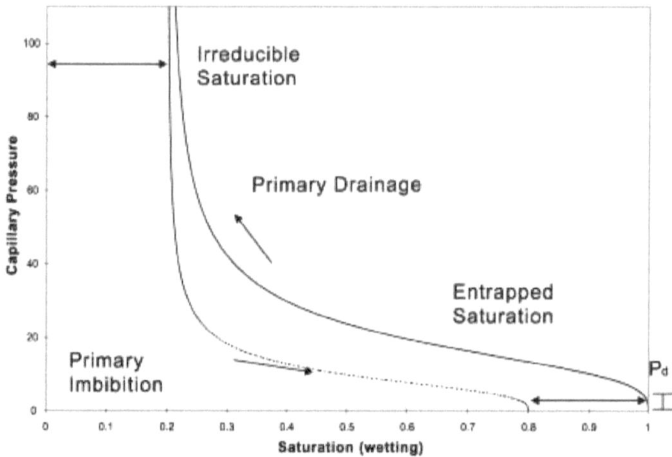

FIGURE 2.8: Primary drainage and imbibition of a partially saturated system [47].

2.4.3 Relative permeability

In the determination of the fluid velocities in partially saturated porous media using *Darcy's* Law, the *Darcy* coefficient κ in Equation (2.1), is expressed as the product of the absolute permeability (intrinsic of the porous medium) and the relative permeability of the fluid phase. The relative permeability is a dimensionless parameter that varies from zero to one and is a function of the degree of saturation [37, 48].

Various authors have developed several relative permeability functions. Lewis and Schrefler [37] made use of the relationship proposed by Brooks and Corey [49] in which the relative permeabilities of the liquid phase, k^{rL} and the gas phase k^{rG} were respectively given as

$$k^{rL} = S_e^{\frac{(2+3\lambda)}{\lambda}} \tag{2.8}$$

$$k^{rG} = (1 - S_e)^2 (1 - S_e^{\frac{(2+3\lambda)}{\lambda}}), \tag{2.9}$$

where

$$S_e = \frac{(s^L - s^{LC})}{(1 - s^{LC})}$$

is the effective saturation, s^{LC} is the irreducible saturation and λ is the pore size distribution index.

Gawin et al. [50] made use of the relative gas permeability given by the equation

$$k^{rG} = 1 - \left(\frac{s^L}{s^{cr}} \right)^{A_g},$$ (2.10)

where s^{cr} is the critical degree of saturation beyond which there is no gas flow and A_g is a constant with the range from 1 to 3. The same study presented two different relative permeability functions for the liquid phase. The first function was given as

$$k^{rL} = \left(\frac{s^L - s_{ir}}{1 - s_{ir}} \right)^{A_l}$$ (2.11)

where s_{ir} similar to s^{LC}, is the irreducible saturation and A_l is a constant with the range from 1 to 3. The second function was given as

$$k^{rL} = \left[1 + \left(\frac{1 - RH}{0.25} \right)^{B_l} \right]^{-1} (s^L)^{A_l}$$ (2.12)

where RH is the relative humidity and B_l is a constant either equal to 6 or 16. Equation (2.12) has the advantage over Equation (2.11) of having better numerical properties since it does not make use of the irreducible saturation which has produced theoretical and numerical difficulties in previous works [50].

The relative permeability functions used in this work were obtained from Monlouis-Bonnaire et al. [48]. These were based on van Genuchten [46] relations, with the empirical parameters calibrated for cement-based materials.

The relative permeability of the liquid phase was given by the equation

$$k^{rL} = (s^L)^p \left[1 - \left(1 - (s^L)^{\frac{1}{m}} \right)^m \right]^2$$ (2.13)

and for the gas phase as

$$k^{rG} = (1 - s^L)^p \left[1 - (s^L)^{\frac{1}{m}} \right]^{2m}$$ (2.14)

where m was determined to be 0.56, and p was determined to be 5.5.

2.4.4   Moisture potential function

The moisture potential function of a cementitious porous medium is used to distinguish the range in which the sorption-desorption isotherm is measurable. This depends on whether the saturation degree falls within or outside the hygroscopic moisture range [51].

The hygroscopic moisture range shown in Figure 2.9 was defined by Gawin et al. [50] to be the degree of saturation less than or equal to the solid saturation point. In this range, the liquid phase exists as only adsorbed water and therefore does not contribute a liquid pressure. Beyond the solid saturation point is the capillary region in which both bound and capillary water are present. Only capillary water in the capillary region has mobility.

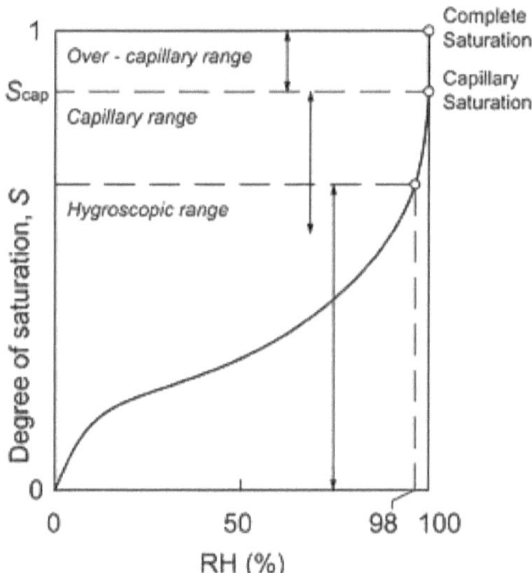

FIGURE 2.9: Hygroscopic and capillary moisture ranges for concrete [52, 53].

At capillary saturation, the capillary pores are said to be saturated with the wetting fluid. However above capillary saturation, the entrained air voids are still not filled with liquid. This results in the difference observed between complete and capillary saturation known as the over-capillary moisture range. In this range there is no moisture gradient, hence very little to no absorption of the liquid into the entrained air voids occurs [53].

The sorption-desorption isotherm described previously is only valid above the hygroscopic moisture range and therefore capillary pressure loses its significance for saturations below the solid saturation point. This limitation makes it necessary to make use of the water potential to describe the moisture state of the medium below the solid saturation point [20, 51].

The water potential Ψ_C is the potential energy of water in a porous medium relative to pure water in the fully saturated condition known as the reference state [54]. It expresses the tendency of water to advance from one region to another as a result of gravity, pressure or capillary suction. For temperatures below the critical temperature of water as employed in this study, the water potential is given by the equation

$$\Psi_C = \frac{RT}{M_w} \ln \left(\frac{p^{GL}}{p^{GLS}} \right),$$

(2.15)

where p^{GL} is the partial pressure of the water vapour, p^{GLS} is the saturation pressure of the water vapour, R is the universal gas constant, T is the temperature and M_w is the molar mass of water [20, 55]. For capillary water in equilibrium with water vapour above a meniscus, the *Kelvin* equation in the form

$$ln \left(\frac{p^{GL}}{p^{GLS}} \right) = -\frac{p^C}{\rho^L} \frac{M_w}{RT}$$

(2.16)

is substituted into Equation (2.15) to obtain the following relationship between capillary pressure and water potential

$$p^C = -\Psi \rho^L.$$

(2.17)

Gawin et al. [20] used Equation (2.17) to justify the use of capillary pressure as a moisture state variable in the simulation of their multiphase model for saturation degrees outside the hygroscopic moisture range. They cautioned that the results obtained in this case could not be interpreted to be pressure but rather a thermodynamic potential for the adsorbed water.

2.5   Chloride-induced reinforcing steel corrosion

Montemor et al. [29] defines corrosion as an electro-chemical process in which chemical reactions generate electrical energy. These reactions occur at anodic and cathodic sites created as a result of chemical or physical variations on the steel surface [56]. The variations are typically borne of the steel manufacturing process or by varying oxygen concentrations at the steel surface [5]. Before corrosion initiation, the anodic reaction is given by the equation

$$Fe \rightarrow Fe_{2+} + 2e^- \tag{2.18}$$

and the balancing cathodic reaction by

$$\frac{1}{2}O_2 + H_2O + 2e^- \rightarrow 2OH^-. \tag{2.19}$$

Although reinforcing steel is naturally subject to corrosion in the atmosphere, the high alkalinity of the cement matrix in which the steel is embedded protects it from corroding [9]. The high pH coupled with the products of both the cathodic and anodic reactions results in the formation of a thin layer of oxide or hydroxide that passivates the steel reinforcing [5, 27].

Corrosion agents do not have a significant effect on the concrete matrix alone. They attack the passive layer of oxide on the steel and actuate corrosion as elaborated in Section 2.5.1 [5]. The concrete cover protects the steel by preventing the ingress of corrosion agents that initiate and sustain corrosion [9]. Apart from chlorides in the marine environment and de-icing salts, chloride ions are introduced into concrete during mixing as a contaminant or in an admixture.

When chloride ingress occurs in a concrete already affected by carbonation, the dual action of the two phenomena increases the rate of deterioration of the reinforcing steel. The carbonation reaction lowers the pH of the concrete matrix thereby allowing the chlorides bound by the products of cement hydration to be released. The ensuing increase in the amount of free chloride ions available to the steel surface results in accelerated corrosion of the reinforcement [5].

Cracking of the cover concrete occurs when the stresses due to rust build-up at the steel-concrete interface exceed the tensile strength of concrete. The cracks then begin to migrate and are halted in regions where the concrete is able to sustain these stresses [26]. Cracks in reinforced concrete are known to increase the penetrability of the concrete and hence significantly increase the rate of steel corrosion. The degree to which the corrosion rate is affected depends on the crack properties such as crack width, frequency and density [9].

Liu and Weyers [13] found that a proportion of the corrosion products fill the porous zone of hydrated cement paste adjacent to the steel. This means that only part of the rust formed during corrosion contributes to the expansive stresses that result in cover cracking. Therefore when developing time-to-cracking models, it is vital that this behaviour is accounted for in order to make an accurate determination of service life.

2.5.1   Depassivation of steel

Revie and Uhlig [56] define passivation as the state in which a metal known to thermodynamically react in a certain environment resists significant corrosion.

Under normal conditions, the high alkalinity of concrete provides protection to the steel from corrosion by passivation. If a concrete has low water/cement ratio, is well compacted and appropriately cured, it's low permeability should minimize the ingress of corrosion-inducing agents. Concrete is also known to have low electric conductivity which reduces the flow of electric current from the anode to the cathode, thereby reducing the corrosion rate [57].

All these properties mean that in an ideal situation where the reinforced concrete is well designed, cast and maintained, reinforcement corrosion should not be a major cause of deterioration of the structure. However, this is rarely the case in reality [57].

Depassivation may occur due to chlorides present in the concrete at casting. Alternatively it may occur over a significant period of time after construction due to ingress of chlorides or diffusion of carbon dioxide in air into concrete, causing the conversion of alkaline $Ca(OH)_2$ to the less alkaline $CaCO_3$ [56].

2.5.2   The corrosion cell

Reinforcement corrosion was previously established to be an electrochemical process [29]. The coupled anodic and cathodic reactions characteristic of the process take place on the surface of the steel which also acts as the electrical link between the two. The pore water in the concrete acts as the electrolyte and the whole system creates what is known as a corrosion cell [57]. Current flows between the anode and cathode of the cell due to the potential difference between the two electrodes [9].

Corrosion cannot occur if the steel is passivated as described in Section 2.5.1; if there is no oxygen available at the cathode or if the concrete is completely dry [26].

The half cell reactions

The anodic and cathodic reactions are collectively known as the half cell reactions. These are reduction-oxidation reactions with the oxidation reaction taking place at the anode and the reduction process taking place at the cathode as shown in Figure 2.10. At the anode, iron (Fe) is dissolved into solution while at the cathode oxygen (O_2) is reduced to form hydroxyl ions (OH^-) [57, 56].

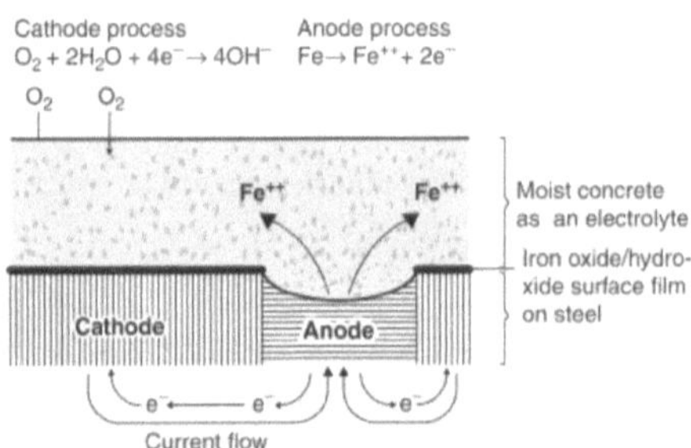

FIGURE 2.10: Corrosion of reinforcement steel bars in concrete [4].

The oxidation reaction of iron at the anode is given in Equation (2.18). The Fe^{2+} produced continues to react with hydroxyl ions to produce the final corrosion products which will collectively be referred to as rust. These reactions depend on the pH of the electrolyte (pore water) and the concentration of the aggressive ions and are given by the equations [12]

$$Fe^{2+} + 2OH^- \rightarrow Fe(OH)_2 \tag{2.20}$$

$$4Fe(OH)_2 + 2H_2O + O_2 \rightarrow 4Fe(OH)_3 \tag{2.21}$$

$$2Fe(OH)_3 \rightarrow 2H_2O + Fe_2O_3 \cdot H_2 0. \tag{2.22}$$

Chloride ions are not consumed in the production of rust but act as catalysts according to the reversible reaction [9]

$$2Fe^{2+} + 6Cl^- \leftrightarrow 2FeCl_3^- + 4e^- \tag{2.23}$$

$$2FeCl_3^- + 4OH^- \leftrightarrow 2Fe(OH)_2 + 6Cl^-. \tag{2.24}$$

This reaction occurs at the anode and results in chloride ions always being available to sustain corrosion. The reactions at the cathode depend on the concentration of oxygen and on the pH at the steel surface.

2.5.3 Rate of corrosion

Otieno et al. [58] list the principal aspects to take into consideration when developing a corrosion rate prediction model. These aspects are: (1) Corrosion rate varies with the age of the structure and hence is time-variant; (2) Cover cracking significantly influences the corrosion rate; (3) Significant differences have been observed in the measurement of corrosion rate using different instruments and/or techniques under identical conditions; (4) Because most corrosion rate models are developed using accelerated corrosion experimental results, validation of the models is vital in order to ensure the results are representative of natural corrosion; (5) Due to the variability of the corrosion rate, the variability of the input parameters that affect corrosion rate should be analysed beforehand.

Corrosion rate depends on the rate at which iron is dissolved at the anode, the availability of water and oxygen at the cathode and the rate at which hydroxyl ions are transported through the electrolyte. The dissolution of iron at the anode is time-dependent as the thickness of the corrosion layer increases with time. This increase in thickness reduces the rate at which aggressive ions diffuse to reach the steel surface and hence reduce the corrosion rate [26].

Corrosion rate is also significantly affected by the pore structure of the concrete matrix and its degree of saturation. In experimental studies, corrosion rates have been observed to be very low below 50% relative humidity and non-existent below 35% relative humidity. Corrosion rate increases rapidly between 50% and 70% and stabilizes between 70% and 90% before decreasing towards 100% relative humidity [26].

2.5.4 Products of reinforcement corrosion

Reinforcement corrosion products vary significantly depending on the prevailing conditions such as the pH of the pore solution, oxygen availability and moisture content. A general formula for these products may be written as $Fe(OH)_2n.Fe(OH)_3p.H_2O$ [13]. The corrosion layer typically consists of a complex mixture of iron oxy-hydroxides namely Goethite ($\alpha - FeOOH$), Akageneite ($\beta - FeOOH$), Lepidocrocite ($\gamma - FeOOH$), Maghemite ($\gamma - Fe_2O_3$) and Magnetite (Fe_3O_4). The different products have differing densities and volume expansions as shown in Figure 2.11, which create expansive stresses at the steel-concrete interface [26].

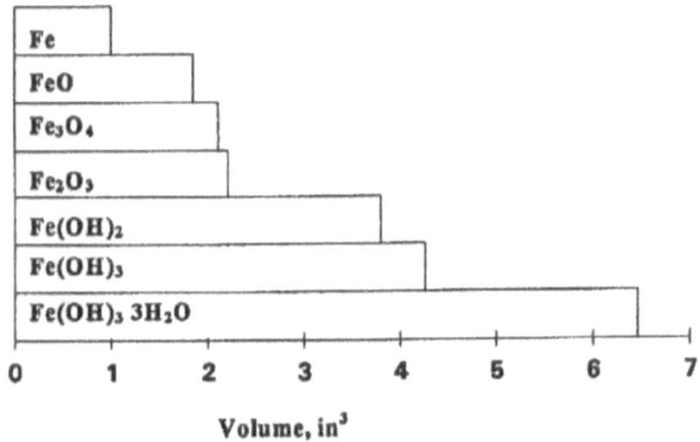

FIGURE 2.11: Relative volume ratios between Iron and the products of corrosion. [13].

Although existing models tend to model the corrosion product as uniformly distributed around the steel bar, it has generally been observed that pitting due to chloride-induced corrosion results in a non-uniform distribution of the corrosion product [3, 36, 59]. In this case, rust tends to accumulate on the side of the rebar with the least cover resulting in the geometry shown in Figure 2.12.

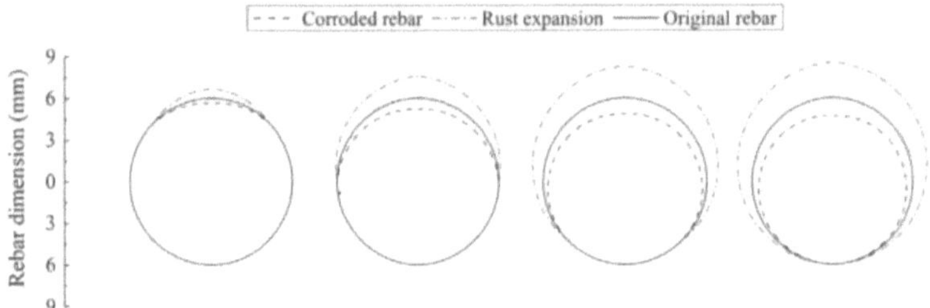

FIGURE 2.12: Rust distribution in pitting corrosion. [59].

2.5.5 Movement of rust in concrete

A theory accounting for the movement of rust into the concrete capillary pores was suggested by Zhao et al. [10]. In their study, it was found that the formation of the corrosion layer and the movement of rust into the porous concrete occur simultaneously after corrosion has been initiated.

Despite the apparent insolubility of the final corrosion products, the process of corrosion involves the formation of unstable chloride complexes with iron as shown in Equation (2.23).

The iron-chloride complexes are soluble species that can dissolve in the concrete pore solution and subsequently diffuse through the cement paste matrix away from the corroding steel surface where there are low oxygen concentrations. When the complexes come into contact with hydroxyl ions in the concrete pores, they are oxidized to form insoluble iron hydroxides [10].

As hydroxyl ions are readily available close to the steel surface during corrosion, the soluble complexes do not have to migrate far before they are oxidized. This means that the concrete pores in the ITZ around the rebar are rapidly filled with rust, making the depth of penetration of the corrosion product quite small as observed by Michel et al. [3].

2.6 Review of rust development and transport models

Various researchers have identified the need to include the penetration of corrosion products into the concrete pores in corrosion initiation and corrosion propagation models [1, 2, 3]. Not all the models have been discussed in this section. The models presented have endeavoured to include the penetration of corrosion products into the concrete pores and micro-cracks, how this process affects corrosion-induced cracking and to determine the size of the corrosion accommodating region. Information about the quantity of corrosion products penetrating into the concrete pores and the size of the CAR are important parameters in developing more accurate time-to-cracking service life models for reinforced concrete structures.

2.6.1 Toongoenthong and Maekawa [1]

Toongoenthong and Maekawa [1] produced a two-phase mechanical model to simulate corrosion-induced cracking with penetration of corrosion products into the corrosion-induced cracks. The corrosion product and intact steel were represented as one unified phase and the surrounding concrete as the other phase. A uniform corrosion product was assumed to be magnetite (Fe_3O_4) and this was assumed to be uniformly distributed around the intact steel. These assumptions were made based on observations of the corrosion products in the accelerated corrosion tests used to verify the model [1].

Non-linear finite element analysis was used in the two dimensional thick-walled cylinder cracking model to estimate the critical mass of steel loss resulting in concrete cover cracking [1]. The mechanism for migration of the corrosion products into the corrosion-induced cracks was attributed to the corrosion product existing as a gel with liquid properties. The ability of the corrosion products to migrate into the corrosion-induced cracks was termed *buffer capacity* (Q_{cr}) and was calculated using the equation

$$Q_{cr} = \oint_{Domain} (\mathbf{u} \cdot \mathbf{n}) \, ds, \tag{2.25}$$

where $\mathbf{u}$ is the displacement vector and $\mathbf{n}$ is the unit vector normal to the boundary surface of the domain in the finite element analysis [1]. Subsequently, the volume loss per unit length of steel resulting in expansive stresses at the steel-concrete interface $V_{effective}$ was calculated as

$$V_{effective} = V_{loss} - Q_{cr}, \tag{2.26}$$

where V_{loss} is the total volume of steel lost per unit length.

Initially, the existence of a buffer capacity was an assumption made due to discrepancies between simulated and experimental results in previous research. The authors performed an experiment to verify the migration of the corrosion products into the corrosion-induced cracks before the cracks migrated to the concrete surface. Figure 2.13 shows a specimen from the aforementioned experiment where a corrosion product front is observed to penetrate into the concrete.

Analyses to determine the critical mass loss with and without the inclusion of buffer capacity were performed to justify its inclusion in the model. The results of the numerical simulations including the buffer capacity were consistent with the experimental data used for the verification [1].

The inclusion of creep deformations in the determination of the critical steel mass loss was found to be vital especially for reinforced concrete members with a high cover depth to reinforcement diameter ratio and those subjected to long term corrosion. Neglecting to include creep effects resulted in a lesser critical steel mass loss compared to the experimental data. Creep effects were accounted for by using a creep coefficient to effectively reduce the elastic stiffness of the concrete surrounding the steel. This increased the critical steel mass loss before cover cracking, resulting in good agreement with experimental results [1].

Val et al. [2] pointed out that there was no consideration for the penetration of corrosion products into the concrete pore spaces in this model and that Toongoenthong and Maekawa [1] did not make

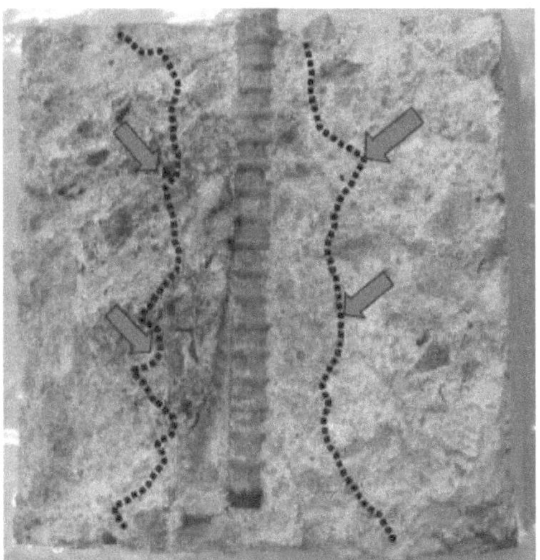

FIGURE 2.13: Migration of the corrosion products into the concrete during experiments performed by Toongoenthong and Maekawa [1].

use of experimental data to study penetration of corrosion products into the concrete cracks during the crack propagation phase.

One of the major findings of this study was that the penetration of the corrosion gel into the corrosion-induced cracks was insignificant for RC specimens with low cover depth to reinforcement diameter ratios. Upon review of these parameters, it was noted that these specimens consistently had cover depths below 50 mm. Such low cover depths result in an increased corrosion rate as the corrosion agents do not have to penetrate through a thick cover. An increased corrosion rate results in a shorter time to cover cracking. In specimens with larger cover depths and with longer times to surface cracking, the penetration of the corrosion gel into the cracks was found to be substantial. This finding corresponds to the finding of Michel et al. [3] that the quantity of corrosion products penetrating into the concrete is directly proportional to the time required for cover cracking.

2.6.2 Val et al. [2]

Val et al. [2] aimed to estimate the amount of corrosion products penetrating into the concrete pores and micro-cracks and to determine the thickness of the porous zone of concrete next to the reinforcing steel. These parameters were required in order to determine what proportion of the total corrosion products formed contributed to the expansive stresses at the steel-concrete interface.

A non-linear finite element analysis scheme was implemented in the finite element software ABAQUS to investigate crack initiation and propagation caused by the expansive corrosion products, initially with no account for the penetration of the corrosion products [2]. Previously, the quantity of corrosion products penetrating into the concrete pores and micro-cracks had not been directly measured experimentally but had been estimated by fitting experimental results to those obtained from numerical models. For this reason, these experimental results could not be used to validate this part of the model presented by Val et al. [2].

The concrete cracking model without penetration of corrosion products was validated using the results obtained by Williamson and Clark [60] in which pressure was applied to holes inside concrete to induce cracking in the concrete. The results obtained showed good correlation with the experimental results when the diameter of the holes was $8mm$ but showed significant discrepancies for $16mm$ diameter holes.

The corrosion propagation part of the model could not be validated quantitatively without accounting for the penetration of corrosion products into the concrete pore spaces and micro-cracks. To validate the model qualitatively, accelerated corrosion tests were conducted using the impressed current method.

The depth of penetration of the corrosion products (p_{corr}) at time t for cases of constant corrosion rate was calculated using *Faraday's* law of electrolysis as

$$p_{corr} = 0.0116 i_{corr}\ t \tag{2.27}$$

where i_{corr} is the corrosion current density. The corrosion penetration was subsequently used to calculate the free expansion of the rebar radius due to corrosion (δ) *from an experimental result* using the equation

$$\delta = \sqrt{r^2 + (\alpha_v - 1)(2r p_{corr} - p_{corr}^2)} - r \tag{2.28}$$

where r is the radius of the rebar and α_v is the volumetric expansion ratio of the corrosion products.

The free increase in the radius (δ) of the reinforcing steel due to expansion of the corrosion product *from finite element analysis* was calculated analogous to thermal expansion using the equation

$$\delta = \alpha_T\ \Delta T r, \tag{2.29}$$

where α_T is the coefficient of thermal expansion, ΔT is the temperature increase and r is the radius of the reinforcing steel [2]. The choice of α_T and ΔT was found to be arbitrary since the material model used was rate independent.

The results from the modelling efforts of Val et al. [2] showed a higher quantity of corrosion products penetrating into the concrete pores than previously published research. Crack impregnation was found to be a gradual process with cracks in larger concrete cover taking longer to fill.

The study could not establish a relationship between the size of the corrosion accommodating region and the water/binder ratio. Because water/binder ratio affects the porosity of the concrete, it was assumed that the size of the CAR would also depend on the water/binder ratio. However, a direct proportionality was identified between the size of the CAR and the time to crack initiation whereby a longer time to cracking resulted in a larger quantity of rust penetration. Unlike Toongoenthong and Maekawa [1], Val et al. [2] found that inclusion of creep effects related to their accelerated corrosion tests did not influence the size of the CAR.

The penetration of the corrosion products was confirmed to be the cause of discrepancies between simulated and experimental results since the tests conducted showed that the discrepancies were not a result of the constitutive modelling of concrete. This finding highlighted the importance of knowing the amount of corrosion products penetrating into the concrete pores in crack propagation models.

2.6.3 Michel et al. [3]

Michel et al. [3] performed experimental studies on the effect of the corrosion current density and water/binder ratio on the initiation and propagation of corrosion-induced cracks, and on the penetration of corrosion products into the corrosion accommodation region (CAR). The experimental set up employed for X-ray attenuation and digital image correlation (DIC) measurements of the accelerated corrosion test is presented in Figure 2.14. DIC measurements were used to determine the deformations between the steel and concrete and X-ray attenuation measurements were used to obtain information on the penetration of the corrosion products [3].

The results from the X-ray attenuation measurements for all the specimens tested showed good correlation with the total amount of corrosion products calculated using *Faraday's* law. They also indicated that the concentration of the corrosion products and the size of the CAR increased with time [3].

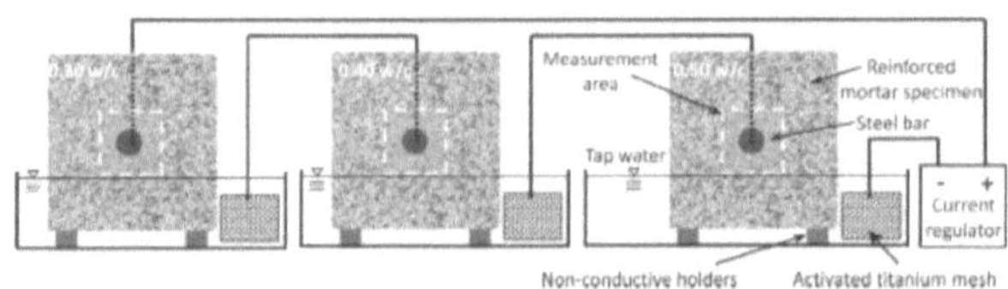

FIGURE 2.14: Experimental set-up for accelerated corrosion testing using the impressed current method by Michel et al. [3].

Further result analysis showed that the maximum size of the CAR was not dependent on the water/binder ratio or the corrosion current density. The lack of influence of the water/binder ratio on the maximum penetration depth was presumed to be due to the discontinuity of the capillary pore network at water/binder ratios less than 0.5 which hinders the movement of the corrosion products. The authors speculated that water/binder ratios greater than 0.6 might influence the maximum size of the CAR [3].

Subsequently Michel et al. [3] developed a combined electrochemical, transport and mechanical numerical model for the simulation of corrosion-induced deformations and cracking. Similarly to Toongoenthong and Maekawa [1], they assumed a uniform corrosion product namely haematite (Fe_2O_3) to be uniformly distributed around the rebar length and perimeter. The assumption was made to allow the use of two dimensional plain strain theory and was later confirmed using X-ray Diffraction and Scanning Electron Microscopy of the corrosion products from the experiments.

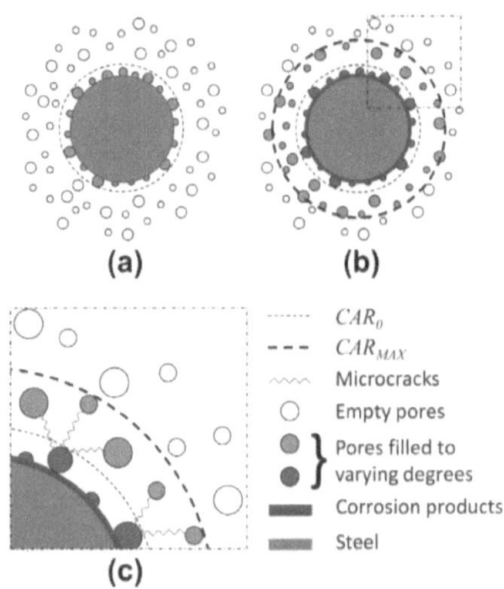

FIGURE 2.15: Expansion of corrosion accommodation region presented by Michel et al. [3].

Figure 2.15 is an illustration of the penetration of corrosion products into the concrete pores. In the figure, CAR_0 is the initial corrosion accommodating region made up of capillary pores which fills up with corrosion products before expansive stresses are induced at the steel-concrete interface. When the CAR_0 is completely filled with corrosion products, micro-cracks begin to appear. The CAR continues to increase until it reaches CAR_{max} at which point no more corrosion products can be accommodated in the capillary pores and increased tensile stresses produce macro-cracks in the concrete cover [3].

The volume of the CAR (V_{CAR}) was calculated using the equation

$$V_{CAR} = \phi \, V_{cm} \tag{2.30}$$

where ϕ is the capillary porosity of the cement matrix and

$$V_{cm} = \pi \, l_A((r + CAR)^2 - r^2) \tag{2.31}$$

is the volume of the matrix accessible to the corrosion products dependent on the size of the CAR. l_A is the length of the anode and r is the radius of the original steel bar. The increase from CAR_0 to CAR_{max} was obtained from the equation

$$CAR = CAR_0 + (CAR_{max} - CAR_0)\kappa \tag{2.32}$$

where κ is a dimensionless coefficient accounting for the change in the connectivity of the capillary pores of the CAR and varies from 0 to 1. The penetration of the corrosion products into V_{CAR} was described by the parameter λ_{CAR} according to

$$\lambda_{CAR} = \begin{cases} \left(\dfrac{V_{cp}}{V_{CAR}}\right)^n, & \text{if } V_{cp} < V_{CAR} \\ 1, & \text{if } V_{cp} \geq V_{CAR} \end{cases} \tag{2.33}$$

where n is a non-physical fitting parameter and V_{cp} is the volume of expanded corrosion products.

Similarly to Val et al. [2], the volumetric expansion of the corrosion products was modelled analogous to thermal loading at the corroding section. However the coefficient of thermal expansion α from Equation (2.29) in Michel et al. [3]'s model was assumed constant and was derived to be a third of the volume expansion coefficient of the corrosion product. An adjusted incremental temperature ΔT_{CAR} was used to include the influence of the CAR on the deformations resulting from corrosion and was given by

$$\Delta T_{CAR} = \lambda_{CAR}\Delta T. \tag{2.34}$$

where ΔT is the incremental temperature as described in Equation (2.29).

The numerical model was implemented in a commercial FEM software and was validated twice; once using the aforementioned experimental results and once using the experimental results of Vu et al. [61]. The simulated results showed good correlation with the experiments for corrosion-induced crack initiation and propagation. However whereas the experiments had revealed non-uniform penetration of the corrosion products, the model produced uniform penetration of the corrosion product into the CAR. The discrepancy led to over or under estimation of the corrosion-induced deformations of the steel-concrete interface at some locations around the rebar [3]. This study found the size of the CAR to be independent of the water/binder ratio and corrosion current density hence confirming the observations of Val et al. [2].

2.6.4   Discussion

From the results of the studies, it is concluded that the corrosion products penetrate into both the concrete pores and the micro-cracks at the steel-concrete interface. The corrosion products fill the corrosion-induced cracks gradually after they are formed. Cracking in the cover concrete occurs before the maximum size of the CAR is attained. Although Val et al. [2] found that the CAR increases with increased time-to-cracking of the concrete surface, the size of the CAR was established to be independent of the corrosion current density and the water/binder ratio and hence the porosity of concrete. This calls into question the use of *Faraday's* law which makes use of corrosion current density and the use of capillary porosity in calculating the volume of corrosion products and the CAR in Equations (2.27) and (2.30).

The three models presented all couple the penetration of the corrosion product with a cracking model of corrosion initiation and propagation. In this work, the penetration of corrosion products is treated in isolation from cracking due to corrosion-induced stresses. This simplification is justifiable at this preliminary stage of the model's development as the complete development of the model is beyond the scope of this work. For the same reason the assumption of one iron-chloro complex ($FeCl_3^-$) is made, analogous to the assumption of a uniform singular corrosion product in all the models.

2.7 The Theory of Porous Media (TPM) and rust transport

Porous media theory is a branch of continuum mechanics that was initially developed for application to soil mechanics. This subject has however advanced to such an extent that it is now applied to various other problems including the modelling of biological tissue [19], cementitious materials [22] and paper processing [62].

The porous nature of the concrete matrix was previously described in Section 2.2. Since the concrete pore spaces are typically saturated with one or more fluids, the whole system may be defined as a multiphasic porous medium as shown in Figure 2.16.

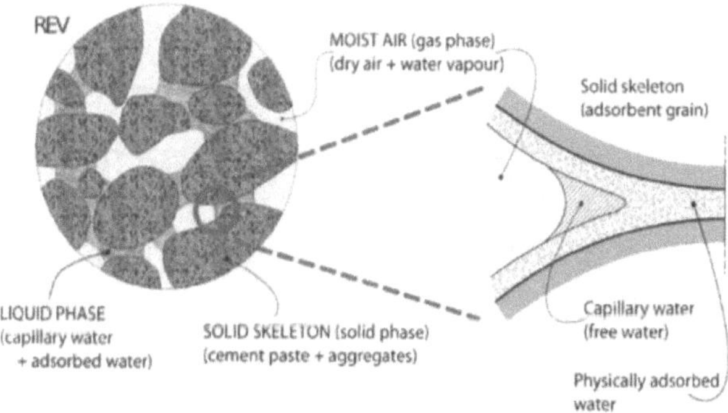

FIGURE 2.16: Schematic of concrete saturated with water and air [22].

The main features of porous media are stipulated in Pesavento et al. [22] as: (1) Each of the phases in the multiphase system may comprise multiple substances; (2) Interactions, mass exchanges and momentum exchanges exist between the various phases; (3) There is mass transport of the fluid constituents unless hygral equilibrium is assumed; (4) The macroscopic behaviour of the porous medium is directly linked to the micro-scale physics of the solid body.

2.7.1 Previous application of TPM to concrete degradation

Existing models of concrete based on porous media theory typically address heat, mass, moisture and solute transport coupled with the mechanical response of the material. The models presented in this section follow a similar procedure in which balance equations are derived from the continuity equation [15]. The balance equations typical of all multiphase models are the balance of mass, the balance of linear momentum, the balance of angular momentum and the balance of energy; elaborated in Chapter 3. These equations are written at the micro-scale and are adapted for each of the constituents of the porous medium. A volume averaging technique is used to translate the equations from the microscale to the macroscale [22].

The constitutive equations relevant to the type of model being developed are thereafter derived by evaluating the entropy inequality in order to ensure the second law of thermodynamics is not violated. The balance equations together with the constitutive equations yield a solvable system. For numerical solution the resulting equation system is discretized in space and time before being implemented in finite element analysis software. This procedure was followed in the development of the numerical model presented in this study.

In consideration of the fact that all the models discussed in this section follow the general procedure outlined above, only their distinguishing characteristics are mentioned here.

Pesavento et al. [22] reviewed the application of porous media theory to modelling of deterioration mechanisms relating to concrete. The generic model they presented is applicable to the hydration of the cement paste, durability assessment and to attack by deleterious species. This model was used to simulate cracking of a concrete beam subjected to variable environmental loading and to simulate alkali-silica reactions under variable moisture conditions. The accuracy of the results obtained demonstrate the potential for porous media mechanics to model the intricate phenomena involved in reinforced concrete degradation.

A porous media model based on Hybrid Mixture Theory was developed by Gawin et al. [20] for the analysis of concrete subjected to high temperatures. The model coupled hygro-thermal degradation with the mechanical response of the material. The model was effective at simulating the effect of externally applied loading on the deterioration of concrete at high temperatures.

Sciumè et al. [23] developed a thermal-hygro-mechanical model for the behaviour of concrete at early age with the degree of reaction and mechanical damage as internal variables. A modified *van Genuchten* equation was used to account for the continued hydration of the cement paste in the simulation of autogenous shrinkage. Their model showed good correlation with experimental results.

A heat and mass transport model for concrete was developed by Salomoni et al. [24]. The model coupled elastoplastic and damage behaviour with creep effects in order to describe concrete subjected to medium and high temperatures. In the model, the hydrated cement paste and the aggregate phase were considered to have differing porosity, permeability and diffusivity.

2.7.2 Rationale for the application of TPM to rust transport

The modelling efforts mentioned in Section 2.6 do not explicitly model the mechanism by which the rust product penetrates into the concrete pores. The influence of fluid transport and change in micro-structure in the CAR as a result of precipitation of rust in the concrete pore spaces on the aforementioned process does not seem to be addressed in these model. Fluid transport affects the poro-elastic material response of reinforced concrete due to the pore pressure. It is therefore deduced that the application of these models is significantly limited to cases identical to those used to calibrate and validate the models.

The Theory of Porous Media makes use of mixture theory in which each of the constituents is assumed to statistically occupy the whole space simultaneously resulting in a smeared model as shown in Figure 2.17 [16]. This means the motion, deformations and other physical and geometrical quantities of each of the constituents may be determined as statistical averages of the real quantities [18, 19]. It is for this reason that porous media theory is well suited to simulating the complex multiphasic micro-structure of reinforced concrete and its deterioration.

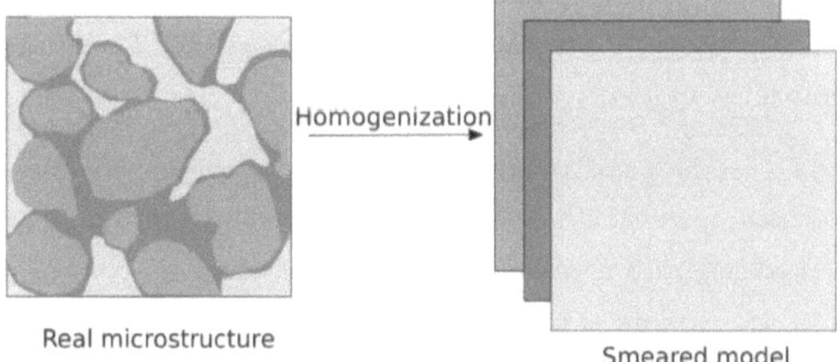

FIGURE 2.17: In TPM, each constituent is assumed to simultaneously occupy the whole control space. Adapted from Hopkins [19].

Using the Theory of Porous Media, the geometrical significance of uniform versus non-uniform corrosion is no longer relevant. Emphasis is rather assigned to the transport of fluids through capillary pores and the pressures these fluids exert on the solid skeleton. With the addition of a concentration into the TPM framework, not only is it possible to model the transport of fluids but the diffusion of ions within the fluids as well. Once this framework is established for one type of solute ion, it may easily be adapted for other ions at sufficiently low concentrations. Consequently, a model designed to simulate the penetration of rust product into the CAR could be adapted to model chloride ingress and carbonation.

2.8 Summary

All deterioration mechanisms of concrete depend on the penetrability of the concrete and its moisture state. These mechanisms depend on the transport of fluids through the concrete pore structure. Reinforcement corrosion is no exception to this fact. Chloride-induced corrosion of the reinforcing steel requires the availability of water, oxygen and chlorides at the steel surface.

In the last two decades inconsistencies have been recorded between time-to-cracking service life models and observations from experimental studies of chloride-induced corrosion in reinforced concrete. Time-to-cracking models generally underestimate the time required before cracks propagate to the concrete surface after corrosion initiation.

The discrepancy has been attributed to not adequately accounting for the penetration of corrosion products into the concrete capillary pores. The corrosion products in the concrete pores do not contribute to the tensile forces at the steel-concrete interface and therefore a larger quantity of steel mass may be lost due to corrosion before cover cracking occurs. This translates into a longer time-to-cracking.

Although efforts have been made to include this process in some time-to-cracking models, no efforts have yet been made to specifically model the penetration of corrosion products. Explicit simulation of the process would yield a more accurate estimate of the quantity of corrosion products in the concrete pores and of the size of the corrosion accommodating region and therefore more accurate time-to-cracking models

The size of the CAR and rate of penetration of the corrosion product were found to be independent of both the corrosion rate and the water/binder ratio of the concrete in previous studies. The rate of penetration decreases as corrosion progresses due to the impregnation of the capillary pores close to the steel surface. It was also determined that the production and penetration of corrosion products into the concrete pores occur simultaneously until the corrosion accommodating region is filled with the rust.

Porous media theory has successfully been applied to a number of deterioration mechanisms of concrete involving heat, mass, moisture and solute transport. However, to the author's knowledge no studies have been done on chloride-induced corrosion of reinforced concrete. In porous media theory, it is not necessary to assume uniform corrosion in the numerical model presented as was done by previous researchers. This is a result of the homogenization process which yields a smeared model in which the exact location of each of the constituents is not required. It is therefore surmised that the Theory of Porous Media may be adapted for modelling chloride-induced corrosion in concrete to yield more accurate time-to-cracking service life models.

3. The Theory of Porous Media

3.1 Introduction

In many disciplines of engineering, it is often necessary to predict the behaviour of a material body in response to internally and externally applied loads. A first step towards this prediction is the description of the physical and chemical composition of the body. Solid bodies may consist of different components and may or may not possess pores filled with fluids. The difference in material properties between the solid and fluid components of a fluid filled porous body results in interactions between the constituents making it difficult to characterize their thermodynamic and mechanical behaviour [16].

Identifying the locations of the aforementioned pore spaces within the solid material to provide a mechanical description of its structure is often complicated. To avoid this problem, the *Theory of Porous Media* (TPM) makes use of mixture theory restricted by the volume fraction concept to describe the dynamic behaviour of macroscopic porous bodies [16, 19]. In this framework, the geometric representation of the pore structure and the exact situation of the individual constituents of the medium need not be regarded [16, 63].

In classical TPM the macroscopic porous body φ consists of κ immiscible phases φ^α with $\alpha = [1, \kappa]$. Each of these phases may exist as a mixture of miscible components ν [18]. From mixture theory, an immiscible phase α containing miscible component ν is referred to as the solvent and the miscible components ν as the solutes [18].

We define the control space as a porous solid B_S with boundary ∂B_S in the actual configuration at time $t \in \mathbb{R}$ and B_{0S} with boundary ∂B_{0S} in the undeformed reference configuration at a fixed time t_0. The whole medium is made up of constituents φ^α with real volumes v^α. The boundary ∂B_S is a material boundary for the solid phases but is a non-material boundary for the liquid and gaseous phases, as they are free to leave the control space [16]. The solid is stipulated to be deformable and therefore the fluid phase and solid phase fields are coupled. Liquid flow occurs as a result of pressure gradients or capillary effects. All the fluids present in the medium are in contact with the solid phase.

3.1.1 Volume fractions

The formulation of the volume fraction concept begins with a theory at the microscale. Continuum mechanical processes naturally occur in the microscopic domain, however the level of detail provided at this scale is often not required for application to practical processes. A representative volume element (RVE) of the microstructure is established so as to allow the expression of the spread of the immiscible components by an averaging volume fraction defined as

$$n^{\alpha}(\mathbf{x},t)[-] = \frac{\mathrm{d}v^{\alpha}[m^3]}{\mathrm{d}v[m^3]}, \tag{3.1}$$

where $\mathbf{x}$ is the position vector in the current configuration, t is the time, $\mathrm{d}v^{\alpha}$ is the element volume of the phase φ^{α} and $\mathrm{d}v$ is the element volume of the bulk medium [16, 18, 19]. The volume fraction concept yields a smeared substitute continuum in which the components are assumed to occupy the whole control space so that the saturation condition may be defined as

$$\sum_{\alpha=1}^{\kappa} n^{\alpha}(\mathbf{x},t) = 1. \tag{3.2}$$

The volume fractions are used to scale the realistic densities of each of the constituents present in the porous medium to obtain to

$$\rho^{\alpha}[g/m^3] = n^{\alpha}\rho^{\alpha R}, \tag{3.3}$$

where ρ^{α} is the partial density and $\rho^{\alpha R}(\mathbf{x},t)$ is the real density of φ^{α}. For the real density of a solution containing a solute suspended in a solvent, the real density is influenced by the molar concentration $c^{\alpha\beta}$ of the solute so that $\rho^{\alpha R} = \rho^{\alpha R}(\mathbf{x},t,c^{\alpha\beta})$ [18].

3.2 Kinematics of porous media

The kinematics of porous media are based on the assumptions that in the actual configuration at time t, each spatial point $\mathbf{x}$ is simultaneously occupied by material points X_{α} of all constituents

φ^α. Each material point follows from a different reference position X_α at time $t = t_0$ and each constituent is assumed to have an independent state of motion [16].

The motion of a constituent φ^α is taken to be a sequence of placements χ_α. Hence for the material points X_α with reference position vector $\mathbf{X}_\alpha$ at time t_0, the spatial position vector $\mathbf{x}$ at time t may be given by [16]

$$\mathbf{x} = \chi_\alpha(\mathbf{X}_\alpha, t). \tag{3.4}$$

The equation above is the Lagrangian description of the motion of the constituents and is a bijective function. χ_α is theorized to be unique and uniquely invertible leading to an inversion of Equation 3.4 to give the Eulerian description of motion for the constituents as

$$\mathbf{X}_\alpha = \chi_\alpha^{-1}(\mathbf{x}, t). \tag{3.5}$$

An assumption is made that the derivatives of the motion functions are continuous. The existence of Equation 3.5 necessitates the definition of the Jacobian as [16]

$$J_\alpha = \det \mathbf{F}_\alpha \tag{3.6}$$

where it is restricted to always be greater than zero to ensure the continuity of the Lagrangian description of motion. $\mathbf{F}_\alpha$ is the deformation gradient defined as [16]

$$\mathbf{F}_\alpha = \frac{\partial \chi_\alpha}{\partial \mathbf{X}_\alpha} = \mathrm{Grad}_\alpha \ \chi_\alpha \tag{3.7}$$

and its inverse by

$$\mathbf{F}_\alpha^{-1} = \frac{\partial \mathbf{X}_\alpha}{\partial \mathbf{x}_\alpha} = \mathrm{grad} -_{,\alpha} \ \mathbf{X}_\alpha \tag{3.8}$$

"Grad" indicates a partial derivative with respect to the reference position $\mathbf{X}_\alpha$ while "grad" indicates a partial derivative with respect to the spatial position $\mathbf{x}$. From Equation 3.4, the Lagrangian velocity and acceleration of a material point are

$$\mathbf{x}_\alpha' = \frac{\partial \chi_\alpha(\mathbf{X}_\alpha, t)}{\partial t} \tag{3.9}$$

$$\mathbf{x}_\alpha'' = \frac{\partial^2 \chi_\alpha(\mathbf{X}_\alpha, t)}{\partial t^2} \tag{3.10}$$

and the Eulerian velocity and acceleration are therefore

$$\mathbf{v}_\alpha = \mathbf{x}'_\alpha = \mathbf{x}'_\alpha(\mathbf{x}, t) \tag{3.11}$$

$$\mathbf{a}_\alpha = \mathbf{x}''_\alpha = \mathbf{x}''_\alpha(\mathbf{x}, t). \tag{3.12}$$

In the *Theory of Porus Media*, the mass averaged velocity relative to the motion of the solid is used to describe the movement of the fluids [37]. The fluid motion is expressed in relation to the current configuration of the solid skeleton.

The individual constituents have independent motion paths therefore independent material time derivatives must be formulated. If $\Gamma(\mathbf{x}, t)$ is an arbitrary scalar-valued and differentiable function, then its material time derivative following the motion of φ^α may be given by [16]

$$\Gamma'_\alpha = \frac{\partial \Gamma}{\partial t} + \operatorname{grad}_{,\alpha} \Gamma \cdot \mathbf{x}'_\alpha \tag{3.13}$$

The material and spatial velocity gradients are respectively given as [16]

$$(\mathbf{F}_\alpha)'_\alpha = \operatorname{Grad}_\alpha \mathbf{x}'_\alpha \tag{3.14}$$

$$\mathbf{L}_\alpha = \operatorname{grad}_{,\alpha} \mathbf{x}'_\alpha \tag{3.15}$$

The spatial velocity gradient describes the deformation process of the solid phase. Upon manipulation of equations 3.14 and 3.15, the spatial velocity gradient may be related to the deformation gradient as

$$\mathbf{L}_\alpha = (\mathbf{F}_\alpha)'_\alpha \mathbf{F}_\alpha^{-1}. \tag{3.16}$$

On additive decomposition of $\mathbf{L}_\alpha$, we obtain a symmetric part $\mathbf{D}_\alpha$ and a skew symmetric part $\mathbf{W}_\alpha$ of the spatial velocity gradient.

$$\mathbf{L}_\alpha = \mathbf{D}_\alpha + \mathbf{W}_\alpha \tag{3.17}$$

where

$$\mathbf{D}_\alpha = \frac{1}{2}(\mathbf{L}_\alpha + \mathbf{L}_\alpha^T) \tag{3.18}$$

$$\mathbf{W}_\alpha = \frac{1}{2}(\mathbf{L}_\alpha - \mathbf{L}_\alpha^T) \tag{3.19}$$

The symmetric part $\mathbf{D}_\alpha$ is defined as the *Eulerian* strain rate tensor associated with pure straining and the skew symmetric part $\mathbf{W}_\alpha$ is the spin tensor.

de Boer [16] stipulates that it is convenient to make use of the difference of the squares of the line elements in both the reference and actual configurations to measure deformations. This is due to the fact that local deformations $\mathbf{F}_\alpha$ contain rigid body motions and are not fit for use in constitutive equations. Usage of the line elements enables the removal of the rigid body motion and circumvents irrational computations. Evaluating the squares of the line-elements;

$$\mathrm{d}\mathbf{x}_\alpha \cdot \mathrm{d}\mathbf{x}_\alpha - \mathrm{d}\mathbf{X}_\alpha \cdot \mathrm{d}\mathbf{X}_\alpha = 2\mathbf{E}_\alpha : (\mathrm{d}\mathbf{X}_\alpha \otimes \mathrm{d}\mathbf{X}_\alpha) \tag{3.20}$$

$$\text{Using} \quad \mathrm{d}\mathbf{x}_\alpha = \mathbf{F}_\alpha \mathrm{d}\mathbf{X}_\alpha, \tag{3.21}$$

$$\mathbf{F}_\alpha \mathrm{d}\mathbf{X}_\alpha \cdot \mathbf{F}_\alpha \mathrm{d}\mathbf{X}_\alpha - \mathrm{d}\mathbf{X}_\alpha \cdot \mathrm{d}\mathbf{X}_\alpha = 2\mathbf{E}_\alpha : (\mathrm{d}\mathbf{X}_\alpha \otimes \mathrm{d}\mathbf{X}_\alpha) \tag{3.22}$$

$$(\mathbf{F}_\alpha^T \mathbf{F}_\alpha - \mathbf{I}) : (\mathrm{d}\mathbf{X}_\alpha \otimes \mathrm{d}\mathbf{X}_\alpha) = 2\mathbf{E}_\alpha : (\mathrm{d}\mathbf{X}_\alpha \otimes \mathrm{d}\mathbf{X}_\alpha) \tag{3.23}$$

$$\therefore \ 2\mathbf{E}_\alpha = \mathbf{F}_\alpha^T \mathbf{F}_\alpha - \mathbf{I} \tag{3.24}$$

$$\mathbf{E}_\alpha = \frac{1}{2}(\mathbf{C}_\alpha - \mathbf{I}) \tag{3.25}$$

where $\mathbf{E}_\alpha$ is the symmetric *Green strain tensor* and $\mathbf{C}_\alpha = \mathbf{F}_\alpha^T \mathbf{F}_\alpha$ is the *right Cauchy-Green deformation tensor*. In a similar manner, the *Almansi strain tensor* $\mathbf{A}_\alpha$ and *left Cauchy-Green deformation tensor* $\mathbf{B}_\alpha$ are defined as [16];

$$\mathrm{d}\mathbf{x}_\alpha \cdot \mathrm{d}\mathbf{x}_\alpha - \mathrm{d}\mathbf{X}_\alpha \cdot \mathrm{d}\mathbf{X}_\alpha = \mathrm{d}\mathbf{x}_\alpha \cdot 2\mathbf{A}_\alpha \mathrm{d}\mathbf{x}_\alpha \tag{3.26}$$

$$\mathbf{A}_\alpha = \frac{1}{2}(\mathbf{I} - \mathbf{B}_\alpha^{-1}) \tag{3.27}$$

$$\text{with} \quad \mathbf{B}_\alpha = \mathbf{F}_\alpha \mathbf{F}_\alpha^T \tag{3.28}$$

3.3   Balance equations

The balance equations are formulated at the microscopic level from classical continuum mechanics for each phase of the porous body and are then transformed to the macroscopic level. This is done by the application of the space averaging technique of integrating the microscopic balance equations over the representative volume element (RVE) to obtain the macroscopic balance equations [21].

3.3.1 Balance of mass

The balance of mass equates the rate of mass to the mass supply as follows

$$(M^\alpha)'_\alpha = \int_{B_\alpha} \hat{\rho}^\alpha \ dv \tag{3.29}$$

with $\hat{\rho}^\alpha$ being the mass supply for φ^α and the mass M^α being given by

$$M^\alpha = \int_{B_\alpha} \rho^\alpha \ dv \tag{3.30}$$

The mass supply is produced by chemical reactions and phase changes between φ^α and the other components occupying the same point at the same time in the current configuration. Using

$$(dv)'_\alpha = \text{div } \mathbf{x}'_\alpha \ dv, \tag{3.31}$$

the rate of mass may be rewritten as

$$(M^\alpha)'_\alpha = \int_{B_\alpha} \left[(\rho^\alpha)'_\alpha + \rho^\alpha \text{div } \mathbf{x}'_\alpha \right] dv \tag{3.32}$$

resulting in the balance of mass being given as

$$\int_{B_\alpha} \left[(\rho^\alpha)'_\alpha + \rho^\alpha \text{div } \mathbf{x}'_\alpha \right] dv = \int_{B_\alpha} \hat{\rho}^\alpha \ dv \tag{3.33}$$

Hence the local form of the balance of mass equation for a phase α is given by

$$(\rho^\alpha)'_\alpha + \rho^\alpha \text{ div } \mathbf{x}'_\alpha = \hat{\rho}^\alpha \tag{3.34}$$

3.3.2 Balance of momentum

The balance of momentum equates the material time derivative of momentum to the sum of external forces as follows [16]

$$(\mathbf{I}_\alpha)'_\alpha = \mathbf{k}_\alpha, \tag{3.35}$$

where $\mathbf{I}_\alpha$ is the momentum of φ^α and is stipulated as

$$\mathbf{I}_\alpha = \int_{B_\alpha} \rho^\alpha \, \mathbf{x}'_\alpha dv \tag{3.36}$$

and $\mathbf{k}_\alpha$ is given by

$$\mathbf{k}_\alpha = \int_{B_\alpha} \rho^\alpha \, \mathbf{b}^\alpha \, dv + \int_{B_\alpha} \hat{\mathbf{p}}^\alpha \, dv + \int_{\partial B_\alpha} \mathbf{t}^\alpha \, da. \tag{3.37}$$

In the above equation, $\rho^\alpha \mathbf{b}^\alpha$ are the volume forces, $\mathbf{t}^\alpha$ are the surface forces and $\hat{\mathbf{p}}^\alpha$ are the interaction forces associated with the volumes [16].

Using *Cauchy's* theorem whereby

$$\mathbf{t}^\alpha = \mathbf{T}^\alpha \mathbf{n}, \tag{3.38}$$

the surface forces may be reformulated as

$$\int_{\partial B_\alpha} \mathbf{t}^\alpha \, da = \int_{\partial B_\alpha} \mathbf{T}^\alpha \mathbf{n} \, da = \int_{\partial B_\alpha} \mathbf{T}^\alpha \, d\mathbf{a} \tag{3.39}$$

where $\mathbf{T}^\alpha$ is the partial *Cauchy* stress tensor of φ^α and $\mathbf{n}$ is the unit normal to the surface of the specific constituent body. We then apply the *divergence theorem*[1] to Equation 3.39 to obtain [16]

$$\int_{\partial B_\alpha} \mathbf{T}^\alpha \mathbf{n} \, da = \int_{B_\alpha} \text{div } \mathbf{T}^\alpha \, dv \tag{3.40}$$

Therefore material time derivative of the momentum $\mathbf{I}_\alpha$ is restated as [16]

$$(\mathbf{I}_\alpha)'_\alpha = \int_{B_\alpha} \rho^\alpha \mathbf{x}''_\alpha + \hat{\rho}^\alpha \mathbf{x}'_\alpha = \int_{B_\alpha} (\text{div } \mathbf{T}^\alpha + \rho^\alpha \, \mathbf{b}^\alpha + \hat{\mathbf{p}}^\alpha) \, dv \tag{3.41}$$

By summing up the above equation over all constituents of the medium we obtain the local form of the balance of momentum for the mixture as [16]

$$\sum_{\alpha=1}^{\kappa} [\text{div } \mathbf{T}^\alpha + \rho^\alpha \, (\mathbf{b}^\alpha - \mathbf{x}''_\alpha)] = \sum_{\alpha=1}^{\kappa} [\hat{\rho}^\alpha \mathbf{x}'_\alpha - \hat{\mathbf{p}}^\alpha] \tag{3.42}$$

---

[1] Also known as *Gauss'* Theorem is given as $\int_\Omega \text{div } \mathbf{q} \, dv = \int_{\partial\Omega} \mathbf{q} \cdot \mathbf{n} \, da$

In comparison to those for a single component medium, the balance of mass and balance of momentum for the mixture must fulfil the following condition [16]

$$\sum_{\alpha=1}^{\kappa} \hat{\mathbf{p}}^{\alpha} = 0 \tag{3.43}$$

3.3.3 Balance of angular momentum

The constituents of the partially saturated porous medium described in this work are considered to be non-polar at the microscopic scale and hence without needing to extensively develop the balance of angular momentum it can be concluded that the partial stress tensor $\mathbf{T}^{\alpha}$ is symmetric, i.e

$$\mathbf{T}^{\alpha} = (\mathbf{T}^{\alpha})^{T} \tag{3.44}$$

and that the summation of the vectors coupling angular momentum between the phases is zero [37].

3.3.4 Balance of energy

The balance of energy is also known as the first law of thermodynamics and equates the sums of the material time derivatives of the internal and kinetic energies to the mechanical work and heat rates. For each constituent in the porous medium, this equation is [16]

$$(E^{\alpha})'_{\alpha} + (K^{\alpha})'_{\alpha} = W^{\alpha} + Q^{\alpha} + \int_{B_{\alpha}} \hat{e}^{\alpha} \, dv, \tag{3.45}$$

where E^{α} is the internal energy, K^{α} the kinetic energy, W^{α} the rate of mechanical work, Q^{α} the rate of heat and $\hat{e}^{\alpha}$ is the energy supply from the other constituents to φ^{α}. These are given by

$$E^{\alpha} = \int_{B_{\alpha}} \rho^{\alpha} \varepsilon^{\alpha} \, dv \tag{3.46}$$

$$K^{\alpha} = \int_{B_{\alpha}} \frac{1}{2} \rho^{\alpha} \mathbf{x}'_{\alpha} \cdot \mathbf{x}'_{\alpha} \, dv \tag{3.47}$$

$$W^{\alpha} = \int_{B_{\alpha}} \mathbf{x}'_{\alpha} \cdot \rho^{\alpha} \mathbf{b}^{\alpha} \, dv + \int_{\partial B_{\alpha}} \mathbf{x}'_{\alpha} \cdot \mathbf{t}^{\alpha} \, dv \tag{3.48}$$

$$Q^{\alpha} = \int_{B_{\alpha}} \rho^{\alpha} r^{\alpha} \, dv - \int_{\partial B_{\alpha}} \mathbf{q}^{\alpha} \cdot \mathbf{da} \tag{3.49}$$

where ε^{α} is the specific internal energy, r^{α} is the partial energy source and $\mathbf{q}^{\alpha}$ is the partial heat flux which is negative when leaving the body. By making use of the following relations,

$$(E^\alpha)'_\alpha = \int_{B_\alpha} \left[\rho^\alpha(\varepsilon^\alpha)'_\alpha + \hat{\rho}^\alpha \varepsilon^\alpha\right] dv, \tag{3.50}$$

$$(K^\alpha)'_\alpha = \int_{B_\alpha} (\rho^\alpha \mathbf{x}''_\alpha + \frac{1}{2}\hat{\rho}^\alpha \mathbf{x}'_\alpha) \cdot \mathbf{x}'_\alpha dv, \tag{3.51}$$

$$\int_{\partial B_\alpha} \mathbf{x}'_\alpha \cdot \mathbf{T}^\alpha \mathbf{n} \, da = \int_{B_\alpha} (\text{div } \mathbf{T}^\alpha \cdot \mathbf{x}'_\alpha + \mathbf{T}^\alpha \cdot \mathbf{L}_\alpha) dv, \tag{3.52}$$

$$\int_{\partial B_\alpha} \mathbf{q}^\alpha \cdot \mathbf{n} \, da = \int_{B_\alpha} \text{div } \mathbf{q}^\alpha dv \tag{3.53}$$

the balance of energy may be rewritten as

$$\int_{B_\alpha} \left[\rho^\alpha(\varepsilon^\alpha)'_\alpha + \hat{\rho}^\alpha \varepsilon^\alpha\right] dv + \int_{B_\alpha} (\rho^\alpha \mathbf{x}''_\alpha + \frac{1}{2}\hat{\rho}^\alpha \mathbf{x}'_\alpha) \cdot \mathbf{x}'_\alpha dv = \int_{B_\alpha} \left[(\text{div } \mathbf{T}^\alpha \cdot \mathbf{x}'_\alpha + \rho^\alpha \mathbf{b}^\alpha) + \mathbf{T}^\alpha \cdot \mathbf{L}_\alpha\right] dv$$
$$+ \int_{B_\alpha} (\rho^\alpha r^\alpha - \text{div}\mathbf{q}^\alpha) \, dv + \int_{B_\alpha} \hat{e}^\alpha \, dv \tag{3.54}$$

and in the local form,

$$\rho^\alpha(\varepsilon^\alpha)'_\alpha + \hat{\rho}^\alpha(\varepsilon^\alpha + \frac{1}{2}\mathbf{x}'_\alpha \cdot \mathbf{x}'_\alpha) = \left[\text{div } \mathbf{T}^\alpha \cdot \mathbf{x}'_\alpha + \rho^\alpha(\mathbf{b}^\alpha + \mathbf{x}''_\alpha)\right] \cdot \mathbf{x}'_\alpha$$
$$+ \mathbf{T}^\alpha \cdot \mathbf{L}_\alpha + \rho^\alpha r^\alpha - \text{div}\mathbf{q}^\alpha + \hat{e}^\alpha \tag{3.55}$$

Due to the symmetry of the partial *Cauchy* stress tensor $\mathbf{T}^\alpha$ as expressed in Section 3.3.3, we can substitute $\mathbf{L}_\alpha$ with $\mathbf{D}_\alpha$ and make use of the local balance of momentum in Equation 3.42 to produce an alternative expression for the balance of energy equation as

$$\rho^\alpha(\varepsilon^\alpha)'_\alpha - \mathbf{T}^\alpha \cdot \mathbf{D}_\alpha - \rho^\alpha r^\alpha + \text{div}\mathbf{q}^\alpha = \hat{e}^\alpha - \hat{\mathbf{p}}^\alpha \cdot \mathbf{x}'_\alpha - \hat{\rho}^\alpha(\varepsilon^\alpha - \frac{1}{2}\mathbf{x}'_\alpha \cdot \mathbf{x}'_\alpha) \tag{3.56}$$

and by summing up over all the κ constituents, we obtain

$$\sum_{\alpha=1}^{\kappa} [\rho^\alpha(\varepsilon^\alpha)'_\alpha - \mathbf{T}^\alpha \cdot \mathbf{D}_\alpha - \rho^\alpha r^\alpha + \text{div}\mathbf{q}^\alpha] = \sum_{\alpha=1}^{\kappa} [\hat{e}^\alpha - \hat{\mathbf{p}}^\alpha \cdot \mathbf{x}'_\alpha - \hat{\rho}^\alpha(\varepsilon^\alpha - \frac{1}{2}\mathbf{x}'_\alpha \cdot \mathbf{x}'_\alpha)] \tag{3.57}$$

We compare the above equation to that of a single constituent material and determine that in order to satisfy the third metaphysical principal of *Truesdell*[2], the following condition must be true and

---

[2]This principal states that "the motion of the mixture body is governed by the same equations as those for a single body." [16]

is sufficient;

$$\sum_{\alpha=1}^{\kappa} \hat{e}^{\alpha} = 0. \tag{3.58}$$

3.4 Entropy inequality

The entropy inequality is used to systematically develop constitutive equations in order to obtain a uniform macroscale thermodynamic description of the material behaviour[3]. Using the entropy inequality guarantees that the second law of thermodynamics is not contravened [37]. The entropy H^{α} is given as

$$H^{\alpha} = \int_{B_{\alpha}} \rho^{\alpha} \eta^{\alpha} \, dv \tag{3.59}$$

where η^{α} is the partial specific entropy. The inequality is defined as

$$\sum_{\alpha=1}^{\kappa} (H^{\alpha})'_{\alpha} \geq \sum_{\alpha=1}^{\kappa} \int_{B_{\alpha}} \frac{1}{\theta^{\alpha}} \rho^{\alpha} r^{\alpha} \, dv - \sum_{\alpha=1}^{\kappa} \int_{\partial B_{\alpha}} \frac{1}{\theta^{\alpha}} \mathbf{q}^{\alpha} \cdot \mathbf{da} \tag{3.60}$$

where θ^{α} is the partial temperature. Using the material time derivative of the entropy

$$(H^{\alpha})'_{\alpha} = \int_{B_{\alpha}} [\rho^{\alpha} (\eta^{\alpha})'_{\alpha} + (\rho^{\alpha})'_{\alpha} \eta^{\alpha}] \, dv, \tag{3.61}$$

and the equation

$$\int_{\partial B_{\alpha}} \frac{1}{\theta^{\alpha}} \mathbf{q}^{\alpha} \cdot \mathbf{da} = \int_{B_{\alpha}} \operatorname{div}(\frac{1}{\theta^{\alpha}} \mathbf{q}^{\alpha}) \, dv, \tag{3.62}$$

the entropy inequality becomes

$$\sum_{\alpha=1}^{\kappa} \int_{B_{\alpha}} [\rho^{\alpha} (\eta^{\alpha})'_{\alpha} + (\rho^{\alpha})'_{\alpha} \eta^{\alpha}] \, dv \geq \sum_{\alpha=1}^{\kappa} \int_{B_{\alpha}} [\frac{1}{\theta^{\alpha}} \rho^{\alpha} r^{\alpha} - \operatorname{div}(\frac{1}{\theta^{\alpha}} \mathbf{q}^{\alpha})] \, dv \tag{3.63}$$

with

$$\rho^{\alpha} r^{\alpha} = -\rho^{\alpha} (\varepsilon^{\alpha})'_{\alpha} + \mathbf{T}^{\alpha} \cdot \mathbf{D}_{\alpha} - \operatorname{div} \mathbf{q}^{\alpha} - \hat{e}^{\alpha} + \hat{\mathbf{p}}^{\alpha} \cdot \mathbf{x}'_{\alpha} + \hat{\rho}^{\alpha} (\varepsilon^{\alpha} + \frac{1}{2} \mathbf{x}'_{\alpha} \cdot \mathbf{x}'_{\alpha}), \tag{3.64}$$

---

[3]The use of the entropy inequality was pioneered by Coleman and Noll [64] and used by Gray and Hassanizadeh [65] in the development of constitutive relations for flow in partially saturated soil.

and

$$\text{div}(\frac{1}{\theta^\alpha}\mathbf{q}^\alpha) = -\frac{1}{\theta^\alpha}\,\text{grad}\theta^\alpha \cdot \mathbf{q}^\alpha + \text{div }\mathbf{q}^\alpha. \tag{3.65}$$

With the *Helmholtz* free energy function

$$\psi^\alpha = \epsilon^\alpha - \theta^\alpha\eta^\alpha, \tag{3.66}$$

its derivative

$$(\psi^\alpha)'_\alpha = (\epsilon^\alpha)'_\alpha - (\theta^\alpha)'_\alpha\eta^\alpha - \theta^\alpha(\eta^\alpha)'_\alpha \tag{3.67}$$

and the local balance of energy from Equation 3.55 in mind, the final form of the entropy inequality is rewritten as

$$\sum_{\alpha=1}^{\kappa}\frac{1}{\theta^\alpha}\{-\rho^\alpha[(\psi^\alpha)'_\alpha + (\theta^\alpha)'_\alpha\eta^\alpha] - \hat{\rho}^\alpha(\psi^\alpha - \frac{1}{2}\mathbf{x}'_\alpha \cdot \mathbf{x}'_\alpha)$$
$$+ \mathbf{T}^\alpha \cdot \mathbf{D}_\alpha - \hat{\mathbf{p}}^\alpha \cdot \mathbf{x}'_\alpha - \frac{1}{\theta^\alpha}\mathbf{q}^\alpha \cdot \text{grad}\theta^\alpha + \hat{e}^\alpha\} \geq 0 \tag{3.68}$$

This entropy inequality is conveniently adequate for obtaining restrictions for the constitutive relations in the model and so other expressions of the entropy inequality will not be dealt with in this work.

4. The triphasic material model with a miscible component

4.1 The triphasic porous body

The existing generic two-phase framework for porous media was presented in the previous chapter [16, 18, 19]. In this chapter, the framework developed for modelling the penetration of iron III chloride using TPM in this study is presented. The two-phase theory is compounded to include three immiscible components identified as the solid phase $\varphi^{\mathbf{S}}$, the liquid phase $\varphi^{\mathbf{L}}$ and the gas phase $\varphi^{\mathbf{G}}$.

The liquid and gas phases are collectively referred to as the fluid phase $\varphi^{\mathbf{F}}$ with the liquid phase defined as the wetting fluid and the gas phase the non-wetting fluid. Wetting is the tendency for a fluid to sustain contact with a solid surface due to intermolecular forces between the two phases. For a comprehensive description on wetting, the reader is referred to Bear and Bachmat [15] section 5.1.1. The two fluids are transported through the pore space of the solid body. Each fluid phase is separated from the other phases by a surface boundary and inhabits a well defined portion of the pore space at the microscopic level [15].

A miscible component φ^{LQ} is included in the liquid phase. Due to the presence of a solute within the liquid phase, a distinction needs to be made between components of the same phase which will be denoted in normal case $(\varphi^{L}, \varphi^{LQ})$ and the phase itself which will be denoted in bold case $(\varphi^{\mathbf{L}})$ so that

$$\varphi^{\mathbf{L}} = \varphi^{L} + \varphi^{LQ}. \tag{4.1}$$

To commence the description of the triphasic material model, the pore saturation degree s with the α phase is defined as [21]

$$s^\alpha = \frac{n^\alpha}{n^{\mathbf{F}}}, \quad \alpha = \mathbf{L}, \mathbf{G}. \tag{4.2}$$

where $n^{\mathbf{F}}$ is the volume of pores within the concrete matrix according to

$$n^{\mathbf{F}} = 1 - n^{\mathbf{S}}, \tag{4.3}$$

and $n^{\mathbf{S}}$ is the solid volume fraction. Equation (4.2) may alternatively be expressed as

$$n^\alpha = n^{\mathbf{F}} \, s^\alpha, \quad \alpha = \mathbf{L}, \mathbf{G}, \tag{4.4}$$

so that we obtain the ansatz

$$s^{\mathbf{L}} + s^{\mathbf{G}} = 1 \tag{4.5}$$

where $s^{\mathbf{L}}$ and $s^{\mathbf{G}}$ are the saturation degrees with the liquid and gas phase respectively.

We assume that the temperature differences between the constituents are negligible resulting in local thermodynamic equilibrium. Therefore at a particular point, the partial temperature θ^α and partial heat flux q^α of all the components of the medium is equal [21]. This simplifies the model in that one set of variables is used to describe the thermodynamic state of all the constituents according to

$$\theta^{\mathbf{S}} = \theta^{\mathbf{L}} = \theta^{\mathbf{G}} = \theta \tag{4.6}$$

$$\mathbf{q}^{\mathbf{S}} = \mathbf{q}^{\mathbf{L}} = \mathbf{q}^{\mathbf{G}} = \mathbf{q}. \tag{4.7}$$

From the above equations, it is deduced that the internal energy supplies are all zero,

$$\hat{e}^{\mathbf{S}} = 0, \quad \hat{e}^{\mathbf{L}} = 0, \quad \hat{e}^{\mathbf{G}} = 0. \tag{4.8}$$

The solid and liquid phases are assumed incompressible and consequently it holds for the densities in the deformed and undeformed configurations

$$\rho^{SR} = \rho_{0S}^{SR} = \text{constant}, \quad \text{and} \quad \text{therefore} \quad (\rho^{SR})'_{\mathbf{S}} = (\rho_{0S}^{SR})'_{\mathbf{S}} = 0; \tag{4.9}$$

$$\rho^{LR} = \rho_{0L}^{LR} = \text{constant}, \quad \text{and} \quad \text{therefore} \quad (\rho^{LR})'_{\mathbf{L}} = (\rho_{0L}^{LR})'_{\mathbf{L}} = 0. \tag{4.10}$$

The gaseous phase is modelled as an ideal gas assumed to be pure oxygen.

A quasi-static state is assumed for the motion of all phases so that dynamic effects may be disregarded leading to

$$x''_{\alpha} = 0; \quad \text{for } \alpha = \mathbf{S}, \mathbf{L}, \mathbf{G} \tag{4.11}$$

4.2 Balance equations for a triphasic continuum

The general balance equations for a porous medium are given in Section 3.3. In this section, the balance equations have been adapted for a triphasic porous medium with a miscible component in the liquid phase.

4.2.1 The balance of mass

The balance of mass for the solid phase

Using the general form of the balance of mass given in Equation (3.34), the balance of mass for the solid phase is [16]

$$(\rho^{\mathbf{S}})'_{\mathbf{S}} + \rho^{\mathbf{S}} \operatorname{div} \mathbf{x}'_{\mathbf{S}} = 0. \tag{4.12}$$

Making use of the equation for partial density

$$\rho^{\alpha} = n^{\alpha} \, \rho^{\alpha R}, \tag{4.13}$$

Equation (4.12) becomes

$$(n^{\mathbf{S}} \, \rho^{SR})'_{\mathbf{S}} + n^{\mathbf{S}} \, \rho^{SR} \operatorname{div} \mathbf{x}'_{\mathbf{S}} = 0 \quad \text{or} \tag{4.14}$$

$$\boxed{(1 - n^{\mathbf{F}})(\rho^{SR})'_{\mathbf{S}} - \rho^{SR}(n^{\mathbf{F}})'_{\mathbf{S}} + (1 - n^{\mathbf{F}}) \, \rho^{\mathbf{S}} \operatorname{div} \mathbf{x}'_{\mathbf{S}} = 0.} \tag{4.15}$$

The balance of mass for the gas phase

The local mass balance equation for the gas is [18]

$$(\rho^G)'_G + \rho^G \operatorname{div} \mathbf{x}'_G = 0 \tag{4.16}$$

which may be rewritten using Equation (4.13) as

$$(n^G \rho^{GR})'_G + n^G \rho^{GR} \operatorname{div} \mathbf{x}'_G = 0. \tag{4.17}$$

Substituting for $n^G = n^F s^G$, we arrive at

$$\boxed{(n^F s^G \rho^{GR})'_G + n^F s^G \rho^{GR} \operatorname{div} \mathbf{x}'_G = 0.} \tag{4.18}$$

The balance of mass for the pore liquid

It is important to reiterate that the liquid phase φ^L is made up of a mixture of a solute, φ^{LQ} and a solvent, the water (φ^L) as previously stated in Equation (4.1). Therefore the balance of mass for the solvent is given by

$$(\rho^L)'_L + \rho^L \operatorname{div} \mathbf{x}'_L = 0 \tag{4.19}$$

which may be rewritten using Equation (4.13) as

$$(n^L \rho^{LR})'_L + n^L \rho^{LR} \operatorname{div} \mathbf{x}'_L = 0. \tag{4.20}$$

Substituting for $n^L = n^F s^L$;

$$\boxed{(n^F s^L \rho^{LR})'_L + n^F s^L \rho^{LR} \operatorname{div} \mathbf{x}'_L = 0.} \tag{4.21}$$

The balance of mass for the solute

The kinematics of porous media presented in chapter 3 are applicable to both solvents and solutes. We therefore proceed to establish the balance of mass for the solute similarly as was done for the solute [18]. The balance of mass for the solute is given by

$$(\rho^{LQ})'_{LQ} + \rho^{LQ} \operatorname{div} \mathbf{x}'_{LQ} = \hat{\rho}^{LQ}. \tag{4.22}$$

It is postulated that the solute does not significantly change the volume of the solvent so that is assumed $dv^L = dv^{\mathbf{L}}$. Hence the solute does not posses a unique volume fraction but is assigned a molar concentration according to

$$c^{LQ}[mol/m^3] = \frac{n^Q_{mol}[mol]}{dv^L} \tag{4.23}$$

$$n^Q_{mol} = \frac{m^Q[g]}{M^Q_{mol}[g/mol]} \tag{4.24}$$

$$m^Q(x,t) = \int_{B_{\mathbf{L}}} c^{LQ} M^Q_{mol} \ \ dv^L = \int_{B_{\mathbf{L}}} n^{\mathbf{L}} c^{LQ} M^Q_{mol} \ \ dv \tag{4.25}$$

where c^{LQ} is the concentration of the solute $FeCl_3^-$ dissolved in the pure water, n^Q_{mol} is its molar amount, m^Q is the partial mass and M^Q_{mol} is the molar mass. Comparison of Equation (4.25) of the partial mass of the solute with Equation (3.30), the density of the solute is realized to be

$$\rho^{LQ} = n^{\mathbf{L}} c^{LQ} M^Q_{mol}. \tag{4.26}$$

This is substituted into Equation (4.22) to obtain

$$(n^{\mathbf{L}} c^{LQ} M^Q_{mol})'_{LQ} + n^{\mathbf{L}} c^{LQ} M^Q_{mol} \operatorname{div} \mathbf{x}'_{LQ} = \hat{\rho}^{LQ} \tag{4.27}$$

where with no mass supply, $\hat{\rho}^{LQ} = 0$. Bearing in mind the definition of the material time derivative of a scalar quantity and that the molar mass of the solute is a constant, the balance of mass for the solute is rewritten as

$$(n^{\mathbf{L}} c^{LQ})'_{LQ} + n^{\mathbf{L}} c^{LQ} \operatorname{div} \mathbf{x}'_{LQ} = 0. \tag{4.28}$$

Substituting for $n^{\mathbf{L}} = n^{\mathbf{F}} s^{\mathbf{L}}$, the final form of the balance of mass for the solute becomes

$$\boxed{(n^{\mathbf{F}} s^{\mathbf{L}} c^{LQ})'_{LQ} + n^{\mathbf{F}} s^{\mathbf{L}} c^{LQ} \operatorname{div} \mathbf{x}'_{LQ} = 0.} \tag{4.29}$$

4.2.2 Balance of linear momentum for the triphasic continuum

From equations (3.42) and (3.43) and by neglecting inertial effects, the balance of linear momentum for the whole mixture is given as

$$\boxed{\operatorname{div} \mathbf{T} + \rho \mathbf{b} = 0} \tag{4.30}$$

where

$$\mathbf{T} = \mathbf{T^S} + \mathbf{T}^L + \mathbf{T}^{LQ} + \mathbf{T^G} \tag{4.31}$$

$$\rho = \left(1 - n^{\mathbf{F}}\right) \rho^{\mathbf{S}} + \left(n^{\mathbf{F}}\ s^{\mathbf{L}}\right) \rho^L + \left(n^{\mathbf{F}}\ s^{\mathbf{G}}\right) \rho^{\mathbf{G}} \tag{4.32}$$

and $\mathbf{b}$ is the external body force. From Ehlers and Bluhm [38], Equation (4.30) may be written for each constituent as

$$\mathrm{div}\ \mathbf{T^S} + \left(1 - n^{\mathbf{F}}\right) \rho^{\mathbf{S}} \mathbf{b} - \hat{\mathbf{p}}^{\mathbf{F}} = \mathbf{0} \tag{4.33}$$

$$\mathrm{div}\ \mathbf{T}^L + n^{\mathbf{F}}\ s^{\mathbf{L}}\ \rho^L \mathbf{b} + \hat{\mathbf{p}}^L = \mathbf{0} \tag{4.34}$$

$$\mathrm{div}\ \mathbf{T}^{LQ} + \hat{\mathbf{p}}^{LQ} = \mathbf{0} \tag{4.35}$$

$$\mathrm{div}\ \mathbf{T^G} + n^{\mathbf{F}}\ s^{\mathbf{G}}\ \rho^{\mathbf{G}} \mathbf{b} + \hat{\mathbf{p}}^{\mathbf{G}} = \mathbf{0} \tag{4.36}$$

with the consideration of Equation (4.40). From Equation (4.35) it is noticed that the solute is not subjected to the external body force.

4.2.3 The balance of energy for the triphasic continuum

The thermal behaviour of the mixture may be described by the energy balance relation given in Equation (3.57) taking into consideration the internal energy defined as

$$\boxed{\varepsilon^\alpha = \psi^\alpha + \theta^\alpha \eta^\alpha} \tag{4.37}$$

as well as the constraints of the saturation condition, the mass supply and the interaction forces as respectively listed below

$$n^{\mathbf{S}} + n^{\mathbf{L}} + n^{\mathbf{G}} = 1 \tag{4.38}$$

$$\hat{\rho}^{\mathbf{S}} + \hat{\rho}^L + \hat{\rho}^{LQ} + \hat{\rho}^{\mathbf{G}} = 0 \tag{4.39}$$

$$\hat{\mathbf{p}}^{\mathbf{S}} + \hat{\mathbf{p}}^{\mathbf{F}} = \mathbf{0} \tag{4.40}$$

$$\hat{\mathbf{p}}^{\mathbf{F}} = \hat{\mathbf{p}}^L + \hat{\mathbf{p}}^{LQ} + \hat{\mathbf{p}}^{\mathbf{G}} \tag{4.41}$$

The resulting balance equation is used in the evaluation of the entropy inequality.

$$\sum_{\alpha=1}^{\kappa} \left[\rho^\alpha(\psi^\alpha + \theta^\alpha \eta^\alpha)'_\alpha - \mathbf{T}^\alpha \cdot \mathbf{D}_\alpha - \rho^\alpha r^\alpha + \mathrm{div}\mathbf{q}^\alpha\right] = \sum_{\alpha=1}^{\kappa} \left[\hat{e}^\alpha - \hat{\rho}^\alpha(\psi^\alpha + \theta^\alpha \eta^\alpha - \frac{1}{2}\mathbf{x}'_\alpha \cdot \mathbf{x}'_\alpha)\right] \tag{4.42}$$

4.3 Constitutive theory

The balance equations elaborated on in the previous section are coupled with constitutive relations in order to complete the description of the mechanical behaviour of the mixture. It is important when selecting constitutive equations to make use of quantities that can be experimentally measured and to select models that have been extensively validated. These models are typically obtained from evaluation of the entropy inequality and involve the linearisation of more complex cases [37].

4.3.1 Evaluation of the entropy inequality

The entropy inequality is given by

$$\sum_{\alpha} \frac{1}{\theta} \{ -\rho^{\alpha} \left[(\psi^{\alpha})'_{\alpha} + (\theta^{\alpha})'_{\alpha} \eta^{\alpha} \right] - \hat{\rho}^{\alpha} (\psi^{\alpha} - \frac{1}{2} \mathbf{x}'_{\alpha} \cdot \mathbf{x}'_{\alpha}) + \mathbf{T}^{\alpha} \cdot \mathbf{D}_{\alpha} - \hat{\mathbf{p}}^{\alpha} \cdot \mathbf{x}'_{\alpha}$$
$$- \mathbf{q}^{\alpha} \cdot \operatorname{grad}\theta^{\alpha} + \hat{e}^{\alpha} \} \geq 0 \tag{4.43}$$

Given the saturation condition

$$n^{\mathbf{S}} + n^{\mathbf{L}} + n^{\mathbf{G}} = 1, \tag{4.44}$$

and taking its material time derivative with respect to the solid skeleton yields

$$(n^{\mathbf{S}})'_{\mathbf{S}} + (n^{\mathbf{L}})'_{\mathbf{S}} + (n^{\mathbf{G}})'_{\mathbf{S}} = 0 \tag{4.45}$$

From the material time derivative of a differentiable function in Equation (3.13), the material time derivative of the volume fraction is derived as

$$(n^{\alpha})'_{\alpha} = \frac{\partial n^{\alpha}}{\partial t} + \operatorname{grad}_{,\alpha} n^{\alpha} \, \mathbf{x}'_{\alpha} \tag{4.46}$$
$$\text{and} \quad (n^{\alpha})'_{\beta} = \frac{\partial n^{\alpha}}{\partial t} + \operatorname{grad}_{,\alpha} n^{\alpha} \, \mathbf{x}'_{\beta}. \tag{4.47}$$

Therefore

$$(n^{\alpha})'_{\alpha} - \operatorname{grad}_{,\alpha} n^{\alpha} \, \mathbf{x}'_{\alpha} = (n^{\alpha})'_{\beta} - \operatorname{grad}_{,\alpha} n^{\alpha} \, \mathbf{x}'_{\beta}, \tag{4.48}$$
$$(n^{\alpha})'_{\beta} = (n^{\alpha})'_{\alpha} - \operatorname{grad}_{,\alpha} n^{\alpha} \, (\mathbf{x}'_{\alpha} - \mathbf{x}'_{\beta}) \tag{4.49}$$

and finally

$$(n^{\alpha})'_{\beta} = (n^{\alpha})'_{\alpha} - \mathrm{grad}n^{\alpha} \cdot \mathbf{w}_{\alpha\beta}, \tag{4.50}$$

where the substitution for the velocity of α-phase relative to the β-phase $\mathbf{w}_{\alpha\beta} = \mathbf{x}'_{\alpha} - \mathbf{x}'_{\beta}$ has been made. Recalling that the liquid phase $\varphi^{\mathbf{L}}$ is made up of the miscible components φ^{L} and φ^{LQ} that occupy the same volume fraction $n^{\mathbf{L}}$, Equation (4.45) becomes

$$(n^{\mathbf{S}})'_{\mathbf{S}} + (n^{\mathbf{L}})'_{L} + (n^{\mathbf{G}})'_{\mathbf{G}} - \mathrm{grad}n^{\mathbf{L}} \cdot \mathbf{w}_{LS} - \mathrm{grad}n^{\mathbf{G}} \cdot \mathbf{w}_{GS} = 0. \tag{4.51}$$

From Equation (4.13) for partial density, we obtain an alternative form for the material time derivative of the volume fraction $(n^{\alpha})'_{\alpha}$ as

$$(n^{\alpha})'_{\alpha} = \frac{n^{\alpha}}{\rho^{\alpha}}(\rho^{\alpha})'_{\alpha} - \frac{n^{\alpha}}{\rho^{\alpha R}}(\rho^{\alpha R})'_{\alpha} \quad \text{and} \tag{4.52}$$

$$\frac{n^{\alpha}}{\rho^{\alpha}}(\rho^{\alpha})'_{\alpha} = -n^{\alpha}\mathbf{D}_{\alpha} \cdot \mathbf{I}. \tag{4.53}$$

Noting that for the incompressible components,

$$\frac{n^{\alpha}}{\rho^{\alpha R}}(\rho^{\alpha R})'_{\alpha} = 0$$

Equation (4.51) is reformulated as

$$-n^{\mathbf{S}}\mathbf{D}_{\mathbf{S}} \cdot \mathbf{I} - n^{\mathbf{L}}\mathbf{D}_{L} \cdot \mathbf{I} - n^{\mathbf{G}}\mathbf{D}_{\mathbf{G}} \cdot \mathbf{I} - \frac{n^{\mathbf{G}}}{\rho^{GR}}(\rho^{GR})'_{\mathbf{G}} - \mathrm{grad}\, n^{\mathbf{L}} \cdot \mathbf{w}_{LS}$$
$$- \mathrm{grad}\, n^{\mathbf{G}} \cdot \mathbf{w}_{\mathbf{GS}} = 0. \tag{4.54}$$

Multiplying the above equation with a Lagrange multiplier, λ yields

$$-\lambda n^{\mathbf{S}}\mathbf{D}_{\mathbf{S}} \cdot \mathbf{I} - \lambda n^{\mathbf{L}}\mathbf{D}_{L} \cdot \mathbf{I} - \lambda n^{\mathbf{G}}\mathbf{D}_{\mathbf{G}} \cdot \mathbf{I} - \lambda\frac{n^{\mathbf{G}}}{\rho^{GR}}(\rho^{GR})'_{\mathbf{G}} - \lambda\, \mathrm{grad}\, n^{\mathbf{L}} \cdot \mathbf{w}_{LS}$$
$$- \lambda\, \mathrm{grad}\, n^{\mathbf{G}} \cdot \mathbf{w}_{\mathbf{GS}} = 0. \tag{4.55}$$

From Equation (4.27) the material time derivative of the balance of mass for the solute with respect to the solid skeleton is

$$(n^{\mathbf{L}}c^{LQ})'_{\mathbf{S}} + \mathrm{grad}\, (n^{\mathbf{L}}c^{LQ}) \cdot \mathbf{w}_{LQS} + n^{\mathbf{L}}c^{LQ}\, \mathbf{D}_{LQ} - \hat{\rho}^{LQ} = 0, \tag{4.56}$$

where the substitutions

$$(n^{\mathbf{L}}c^{LQ})'_{LQ} = (n^{\mathbf{L}}c^{LQ})'_{\mathbf{S}} + \text{grad }(n^{\mathbf{L}}c^{LQ}) \cdot \mathbf{w}_{LQ\mathbf{S}}$$

and

$$\text{div } \mathbf{x}'_{LQ} = \mathbf{D}_{LQ} \cdot \mathbf{I}$$

have been made. Making use of equations (4.50), (4.52) and (4.53), Equation (4.56) becomes

$$-c^{LQ}n^{\mathbf{L}}\mathbf{D}_L \cdot \mathbf{I} - c^{LQ}\text{ grad }n^{\mathbf{L}} \cdot \mathbf{w}_{LS} + n^{\mathbf{L}}(c^{LQ})'_{LQ} - n^{\mathbf{L}}\text{ grad }c^{LQ} \cdot \mathbf{w}_{LQS}$$
$$+ \text{grad }(n^{\mathbf{L}}c^{LQ}) \cdot \mathbf{w}_{LQ\mathbf{S}} + n^{\mathbf{L}}c^{LQ}\,\mathbf{D}_{LQ} \cdot \mathbf{I} - \hat{\rho}^{LQ} = 0. \tag{4.57}$$

Another Lagrangian multiplier, λ^{LQ} is applied to the previous equation to yield

$$-\lambda^{LQ}c^{LQ}n^{\mathbf{L}}\mathbf{D}_L \cdot \mathbf{I} - \lambda^{LQ}c^{LQ}\text{ grad }n^{\mathbf{L}} \cdot \mathbf{w}_{LS} + \lambda^{LQ}n^{\mathbf{L}}(c^{LQ})'_{LQ} - \lambda^{LQ}n^{\mathbf{L}}\text{ grad }c^{LQ} \cdot \mathbf{w}_{LQS}$$
$$+ \lambda^{LQ}\text{grad }(n^{\mathbf{L}}c^{LQ}) \cdot \mathbf{w}_{LQ\mathbf{S}} + \lambda^{LQ}n^{\mathbf{L}}c^{LQ}\,\mathbf{D}_{LQ} \cdot \mathbf{I} - \lambda^{LQ}\hat{\rho}^{LQ} = 0. \tag{4.58}$$

Equations (4.55) and (4.58) are added to the extended entropy inequality to give

$$\frac{1}{\theta}\{-\rho^{\mathbf{S}}\left[(\psi^{\mathbf{S}})'_{\mathbf{S}} + (\theta^{\mathbf{S}})'_{\mathbf{S}}\eta^{\mathbf{S}}\right] - \rho^L\left[(\psi^L)'_L + (\theta^L)'_L\eta^L\right] - (n^{\mathbf{L}}c^{LQ}M^Q_{mol})\left[(\psi^{LQ})'_{LQ} + (\theta^{LQ})'_{LQ}\eta^{LQ}\right]$$
$$- \rho^{\mathbf{G}}\left[(\psi^{\mathbf{G}})'_{\mathbf{G}} + (\theta^{\mathbf{G}})'_{\mathbf{G}}\eta^{\mathbf{G}}\right] - \hat{\rho}^{\mathbf{S}}(\psi^{\mathbf{S}} - \frac{1}{2}\mathbf{x}'_{\mathbf{S}} \cdot \mathbf{x}'_{\mathbf{S}}) - \hat{\rho}^L(\psi^L - \frac{1}{2}\mathbf{x}'_L \cdot \mathbf{x}'_L) - \hat{\rho}^{LQ}(\psi^{LQ} - \frac{1}{2}\mathbf{x}'_{LQ} \cdot \mathbf{x}'_{LQ})$$
$$- \hat{\rho}^{\mathbf{G}}(\psi^{\mathbf{G}} - \frac{1}{2}\mathbf{x}'_{\mathbf{G}} \cdot \mathbf{x}'_{\mathbf{G}}) + \mathbf{T}^{\mathbf{S}} \cdot \mathbf{D}_{\mathbf{S}} + \mathbf{T}^L \cdot \mathbf{D}_L + \mathbf{T}^{LQ} \cdot \mathbf{D}_{LQ} + \mathbf{T}^{\mathbf{G}} \cdot \mathbf{D}_{\mathbf{G}} - \hat{\mathbf{p}}^{\mathbf{S}}\ \mathbf{x}'_{\mathbf{S}} - \hat{\mathbf{p}}^L \cdot \mathbf{x}'_L$$
$$- \hat{\mathbf{p}}^{LQ} \cdot \mathbf{x}'_{LQ} - \hat{\mathbf{p}}^{\mathbf{G}} \cdot \mathbf{x}'_{\mathbf{G}} - \mathbf{q} \cdot \text{grad}\theta + \lambda n^{\mathbf{S}}\mathbf{D}_{\mathbf{S}} \cdot \mathbf{I} + \lambda n^{\mathbf{L}}\mathbf{D}_L \cdot \mathbf{I} + \lambda n^{\mathbf{G}}\mathbf{D}_{\mathbf{G}} \cdot \mathbf{I} + \lambda\frac{n^{\mathbf{G}}}{\rho^{GR}}(\rho^{GR})'_{\mathbf{G}}$$
$$+ \lambda \text{ grad }n^{\mathbf{L}} \cdot \mathbf{w}_{LS} + \lambda \text{ grad }n^{\mathbf{G}} \cdot \mathbf{w}_{GS} - \lambda^{LQ}c^{LQ}n^{\mathbf{L}}\mathbf{D}_L \cdot \mathbf{I} - \lambda^{LQ}c^{LQ}\text{ grad }n^{\mathbf{L}} \cdot \mathbf{w}_{LS} + \lambda^{LQ}n^{\mathbf{L}}(c^{LQ})'_{LQ}$$
$$- \lambda^{LQ}n^{\mathbf{L}}\text{ grad }c^{LQ} \cdot \mathbf{w}_{LQS} + \lambda^{LQ}\text{grad }(n^{\mathbf{L}}c^{LQ}) \cdot \mathbf{w}_{LQ\mathbf{S}} + \lambda^{LQ}n^{\mathbf{L}}c^{LQ}\,\mathbf{D}_{LQ} \cdot \mathbf{I} - \lambda^{LQ}\hat{\rho}^{LQ}\} \geq 0.$$
$$\tag{4.59}$$

The following assumptions are made for the *Helmholtz* free energy functions

$$\psi^{\mathbf{S}} = \hat{\psi}^{\mathbf{S}}(\mathbf{C}_{\mathbf{S}}) \tag{4.60}$$

$$\psi^L = \hat{\psi}^L(-) \tag{4.61}$$

$$\psi^{LQ} = \hat{\psi}^{LQ}(c^{LQ}) \tag{4.62}$$

$$\psi^{\mathbf{G}} = \hat{\psi}^{\mathbf{G}}(\rho^{GR}). \tag{4.63}$$

Taking their material time derivatives gives,

$$(\psi^{\mathbf{S}})'_{\mathbf{S}} = 2\,\mathbf{F_S}\frac{\partial\psi^{\mathbf{S}}}{\partial\mathbf{C_S}}\mathbf{F}_{\mathbf{S}}^{T}\cdot\mathbf{D_S} \tag{4.64}$$

$$(\psi^{L})'_{L} = - \tag{4.65}$$

$$(\psi^{LQ})'_{LQ} = \frac{\partial\psi^{LQ}}{\partial c^{LQ}}(c^{LQ})'_{LQ} \tag{4.66}$$

$$(\psi^{\mathbf{G}})'_{\mathbf{G}} = \frac{\partial\psi^{\mathbf{G}}}{\partial\rho^{GR}}(\rho^{GR})'_{\mathbf{G}}. \tag{4.67}$$

For the isothermal deformations considered in this work, $(\theta^{\alpha})'_{\alpha}$ and $\operatorname{grad}\theta = 0$ [16]. Making use of the Equations (4.6)-(4.11), along with

$$-\hat{\mathbf{p}}^{\mathbf{S}} = \hat{\mathbf{p}}^{\mathbf{F}} = \hat{\mathbf{p}}^{L} + \hat{\mathbf{p}}^{LQ} + \hat{\mathbf{p}}^{\mathbf{G}} \tag{4.68}$$

and substituting equations (4.64) - (4.67) into Equation (4.59) yields

$$\begin{aligned}
&-2\,\rho^{\mathbf{S}}\mathbf{F_S}\frac{\partial\psi^{\mathbf{S}}}{\partial\mathbf{C_S}}\mathbf{F}_{\mathbf{S}}^{T}\cdot\mathbf{D_S} - (n^{\mathbf{L}}c^{LQ})\frac{\partial\psi^{LQ}}{\partial c^{LQ}}(c^{LQ})'_{LQ} - \rho^{\mathbf{G}}\frac{\partial\psi^{\mathbf{G}}}{\partial\rho^{GR}}(\rho^{GR})'_{\mathbf{G}} - \hat{\rho}^{LQ}(\psi^{LQ})\\
&+\mathbf{T^S}\cdot\mathbf{D_S} + \mathbf{T}^{L}\cdot\mathbf{D}_{L} + \mathbf{T}^{LQ}\cdot\mathbf{D}_{LQ} + \mathbf{T^G}\cdot\mathbf{D_G} - \hat{\mathbf{p}}^{L}\cdot\mathbf{w}_{LS} - \hat{\mathbf{p}}^{LQ}\cdot\mathbf{w}_{LQS} - \hat{\mathbf{p}}^{\mathbf{G}}\cdot\mathbf{w_{GS}}\\
&+\lambda n^{\mathbf{S}}\mathbf{D_S}\cdot\mathbf{I} + \lambda n^{\mathbf{L}}\mathbf{D}_{L}\cdot\mathbf{I} + \lambda n^{\mathbf{G}}\mathbf{D_G}\cdot\mathbf{I} + \lambda\frac{n^{\mathbf{G}}}{\rho^{GR}}(\rho^{GR})'_{\mathbf{G}} + \lambda\operatorname{grad}n^{\mathbf{L}}\cdot\mathbf{w}_{LS} + \lambda\operatorname{grad}n^{\mathbf{G}}\cdot\mathbf{w_{GS}}\\
&-\lambda^{LQ}c^{LQ}n^{\mathbf{L}}\mathbf{D}_{L}\cdot\mathbf{I} - \lambda^{LQ}c^{LQ}\operatorname{grad}n^{\mathbf{L}}\cdot\mathbf{w}_{LS} + \lambda^{LQ}n^{\mathbf{L}}(c^{LQ})'_{LQ} - \lambda^{LQ}n^{\mathbf{L}}\operatorname{grad}c^{LQ}\cdot\mathbf{w}_{LQS}\\
&+\lambda^{LQ}\operatorname{grad}(n^{\mathbf{L}}c^{LQ})\cdot\mathbf{w}_{LQS} + \lambda^{LQ}n^{\mathbf{L}}c^{LQ}\,\mathbf{D}_{LQ}\cdot\mathbf{I} - \lambda^{LQ}\hat{\rho}^{LQ} \geq 0
\end{aligned} \tag{4.69}$$

which may be rewritten as

$$
(\rho^{GR})'_{\mathbf{G}}\{-\rho^{\mathbf{G}}\frac{\partial\psi^{\mathbf{G}}}{\partial\rho^{GR}} + \lambda\frac{n^{\mathbf{G}}}{\rho^{GR}}\}
$$

$$
+ (c^{LQ})'_{LQ}\{-(n^{\mathbf{L}}c^{LQ})\frac{\partial\psi^{LQ}}{\partial c^{LQ}} + \lambda^{LQ}\ n^{\mathbf{L}}\}
$$

$$
+ \mathbf{D_S}\{\mathbf{T^S} - 2\ \rho^{\mathbf{S}}\mathbf{F_S}\frac{\partial\psi^{\mathbf{S}}}{\partial\mathbf{C_S}}\mathbf{F}_{\mathbf{S}}^{T} + \lambda n^{\mathbf{S}}\mathbf{I}\}
$$

$$
+ \mathbf{D}_{L}\{\mathbf{T}^{L} + \lambda n^{L}\mathbf{I} - \lambda^{LQ}\ n^{\mathbf{L}}c^{LQ}\mathbf{I}\}
$$

$$
+ \mathbf{D}_{LQ}\{\mathbf{T}^{LQ} + \lambda^{LQ}\ n^{\mathbf{L}}c^{LQ}\mathbf{I}\} \tag{4.70}
$$

$$
+ \mathbf{D_G}\{\mathbf{T^G} + \lambda n^{\mathbf{G}}\mathbf{I}\}
$$

$$
+ \mathbf{w}_{LS}\{-\hat{\mathbf{p}}^{L} + \lambda\ \mathrm{grad} n^{\mathbf{L}} - \lambda^{LQ}\ c^{LQ}\mathrm{grad}\ n^{\mathbf{L}}\}
$$

$$
+ \mathbf{w}_{LQS}\{-\hat{\mathbf{p}}^{LQ} - \lambda^{LQ}\ n^{\mathbf{L}}\ \mathrm{grad}\ c^{LQ} + \lambda^{LQ}\mathrm{grad}\ (n^{\mathbf{L}}c^{LQ})\}
$$

$$
+ \mathbf{w_{GS}}\{-\hat{\mathbf{p}}^{\mathbf{G}} + \lambda\ \mathrm{grad}\ n^{\mathbf{G}}\}
$$

$$
- \hat{\rho}^{LQ}\{\lambda^{LQ} + \psi^{LQ}\} \geq 0.
$$

Coleman and Noll [64] stipulates that the entropy inequality must hold for fixed values of the process variables

$$
\mathcal{P} = \{\mathbf{C_S}, \rho^{\mathbf{GR}}, c^{LQ}, \hat{\rho}^{LQ}, \mathbf{w}_{LS}, \mathbf{w}_{LQS}, \mathbf{w_{GS}}\}
$$

and for arbitrary values of the freely available variables

$$
\mathcal{A} = \{\mathbf{D_S}, \mathbf{D}_{L}, \mathbf{D}_{LQ}, \mathbf{D_G}, (\rho^{\mathbf{G}R})'_{\mathbf{G}}, (c^{LQ})'_{LQ}\}
$$

which are the derivatives process variables in time and space [18, 38]. Without mass exchange, $\hat{\rho}^{LQ} = 0$. Therefore, the following relations are obtained which are sufficient for the satisfaction of the second law of thermodynamics for the ternary model with no mass supply of the solute $FeCl_3^{-}$

$$\mathbf{T^S} - 2\,\rho^\mathbf{S}\mathbf{F_S}\frac{\partial\psi^\mathbf{S}}{\partial\mathbf{C_S}}\mathbf{F_S^T} + \lambda\,n^\mathbf{S}\mathbf{I} = \mathbf{0}$$

$$\mathbf{T}^L + \lambda n^L\mathbf{I} - \lambda^{LQ}\,n^\mathbf{L}c^{LQ}\mathbf{I} = \mathbf{0}$$

$$\mathbf{T}^{LQ} + \lambda^{LQ}\,n^\mathbf{L}c^{LQ}\mathbf{I} = \mathbf{0}$$

$$\mathbf{T^G} + \lambda n^\mathbf{G}\mathbf{I} = \mathbf{0} \qquad\qquad\qquad (4.71)$$

$$\hat{\mathbf{p}}^L = \lambda\,\mathrm{grad}\,n^\mathbf{L} - \lambda^{LQ}\,c^{LQ}\mathrm{grad}\,n^\mathbf{L}$$

$$\hat{\mathbf{p}}^{LQ} = -\lambda^{LQ}\,n^\mathbf{L}\,\mathrm{grad}\,c^{LQ} + \lambda^{LQ}\mathrm{grad}\,(n^\mathbf{L}c^{LQ})$$

$$\hat{\mathbf{p}}^\mathbf{G} = \lambda\mathrm{grad}\,n^\mathbf{G}$$

with the dissipation

$$\begin{aligned}
\mathcal{D} =&\,\mathbf{w}_{LS}\{-\hat{\mathbf{p}}^L + \lambda\mathrm{grad}\,n^\mathbf{L} - \lambda^{LQ}\,c^{LQ}\mathrm{grad}\,n^\mathbf{L}\} + \mathbf{w}_{LQS}\{-\hat{\mathbf{p}}^{LQ} - \lambda^{LQ}\,n^\mathbf{L}\mathrm{grad}\,c^{LQ} \\
&+ \lambda^{LQ}\mathrm{grad}\,(n^\mathbf{L}c^{LQ})\} + \mathbf{w_{GS}}\{-\hat{\mathbf{p}}^\mathbf{G} + \lambda\,\mathrm{grad}\,n^\mathbf{G}\} \geq 0.
\end{aligned} \qquad (4.72)$$

4.3.2   Stresses

The total Cauchy stress is given by

$$\boxed{\mathbf{T} = \mathbf{T^S} + \mathbf{T}^L + \mathbf{T}^{LQ} + \mathbf{T^G}} \qquad\qquad (4.73)$$

Substituting for

$$\mathbf{T^S} = 2\,\rho^\mathbf{S}\mathbf{F_S}\frac{\partial\psi^\mathbf{S}}{\partial\mathbf{C_S}}\mathbf{F_S^T} - \lambda\,n^\mathbf{S}\mathbf{I} \qquad\qquad (4.74)$$

$$\mathbf{T}^L = -\lambda n^\mathbf{L}\mathbf{I} + \lambda^{LQ}\,n^\mathbf{L}c^{LQ}\mathbf{I} \qquad\qquad (4.75)$$

$$\mathbf{T}^{LQ} = -\lambda^{LQ}\,n^\mathbf{L}c^{LQ}\mathbf{I} \qquad\qquad (4.76)$$

$$\mathbf{T^G} = -\lambda n^\mathbf{G}\mathbf{I} \qquad\qquad (4.77)$$

into Equation (4.73) yields

$$\mathbf{T} = 2\,\rho^S\mathbf{F_S}\frac{\partial\psi^S}{\partial\mathbf{C_S}}\mathbf{F_S^T} - \lambda\,n^\mathbf{S}\mathbf{I} - \lambda\,n^\mathbf{L}\mathbf{I} + \lambda^{LQ}n^\mathbf{L}c^{LQ}\mathbf{I} - \lambda^{LQ}n^\mathbf{L}c^{LQ}\mathbf{I} - \lambda\,n^\mathbf{G}\mathbf{I} \quad \text{and} \qquad (4.78)$$

$$\mathbf{T} = 2\,\rho^S\mathbf{F_S}\frac{\partial\psi^S}{\partial\mathbf{C_S}}\mathbf{F_S^T} - \lambda(n^\mathbf{S} + n^\mathbf{L} + n^\mathbf{G})\mathbf{I}. \qquad\qquad (4.79)$$

Making use of the saturation condition in Equation (4.38) yields the total *Cauchy* stress as

$$\boxed{\mathbf{T} = \mathbf{T}_E^{\mathbf{S}} - p^{\mathbf{S}}\mathbf{I}} \tag{4.80}$$

where

$$\mathbf{T}_E^{\mathbf{S}} = 2\,\rho^S\mathbf{F_S}\frac{\partial\psi^S}{\partial\mathbf{C_S}}\mathbf{F}_{\mathbf{S}}^T \tag{4.81}$$

is the effective stress tensor and $\lambda = p^{\mathbf{S}}$ is the pore pressure defined as an average of the two fluid pressures [15] according to

$$p^{\mathbf{S}} = s^{\mathbf{L}}\,p^{LR} + s^{\mathbf{G}}\,p^{GR}. \tag{4.82}$$

p^{LR} is the pressure exerted by the liquid and p^{GR} is the pressure exerted by the gas of the solid body. From Equation $(4.70)_1$ we obtain

$$\lambda\,\frac{n^{\mathbf{G}}}{\rho^{GR}} = \rho^{\mathbf{G}}\frac{\partial\psi^{\mathbf{G}}}{\partial\rho^{GR}} = n^{\mathbf{G}}\rho^{GR}\frac{\partial\psi^{\mathbf{G}}}{\partial\rho^{GR}} \tag{4.83}$$

$$\lambda = \frac{\rho^{GR}}{n^{\mathbf{G}}}n^{\mathbf{G}}\rho^{GR}\frac{\partial\psi^{\mathbf{G}}}{\partial\rho^{GR}} \tag{4.84}$$

$$\lambda = (\rho^{GR})^2\frac{\partial\psi^{\mathbf{G}}}{\partial\rho^{GR}} \tag{4.85}$$

and from Equation $(4.70)_2$

$$-(n^{\mathbf{L}}c^{LQ})\frac{\partial\psi^{LQ}}{\partial c^{LQ}} + \lambda^{LQ}n^{\mathbf{L}} = 0 \tag{4.86}$$

$$\lambda^{LQ} = c^{LQ}\frac{\partial\psi^{LQ}}{\partial c^{LQ}}. \tag{4.87}$$

From Ricken et al. [18], it is assumed that the *Helmholtz* free energy for the gas and solute constituents respectively are given by

$$\psi^G = -\frac{1}{\rho^{GR}}\left\{R^G\theta\left[\ln\left(\frac{\rho_0^{GR}}{\rho^{GR}}\right) - 1\right]\right\} \tag{4.88}$$

$$\psi^{LQ} = -\frac{1}{c^{LQ}}\left\{R^G\theta\left[\ln\left(\frac{c^{LQ}}{c_0^{LQ}}\right) + 1\right] + \mu_0^{LQ}\right\} \tag{4.89}$$

where p_0 is the atmospheric pressure, ρ_0^{GR} is the reference gas density, μ^{LQ} is the chemical potential of the solute, c_0^{LQ} is the reference solute concentration and μ_0^{LQ} is the reference solute chemical potential [18]. Substituting Equation (4.89) into Equation (4.76) yields

$$\mathbf{T}^{LQ} = -n^{\mathbf{L}} \left(R^G \theta \ln \left(\frac{c^{LQ}}{c_0^{LQ}} \right) + \mu_0^{LQ} \right) \mathbf{I} = -n^{\mathbf{L}} \mu^{LQ} \mathbf{I} \tag{4.90}$$

with $\mu^{LQ} = R^G \theta \ln \left(\frac{c^{LQ}}{c_0^{LQ}} \right) + \mu_0^{LQ}$ [18]. Equation (4.75) is subsequently reformulated to give

$$\mathbf{T^L} = -n^{\mathbf{L}} p^{LR} \mathbf{I} + n^{\mathbf{L}} \mu^{LQ} \mathbf{I} \tag{4.91}$$

Similarly, Equation (4.88) is substituted into Equation (4.77) to give

$$\mathbf{T^G} = -n^{\mathbf{G}} \left(R^G \theta \ln \left(\frac{\rho_0^{GR}}{\rho^{GR}} \right) + p_0 \right) \mathbf{I} = -n^{\mathbf{G}} p^{GR} \mathbf{I} \tag{4.92}$$

where $p^{GR} = R^G \theta \ln \left(\frac{\rho_0^{GR}}{\rho^{GR}} \right) + p_0$ [18].

4.3.3 Seepage velocities for the fluids

To evaluate the seepage velocity for the gas, use is made of the local balance of linear momentum for the gas phase given in Equation (4.36). The *Cauchy* stress for the gas phase from Equation (4.92) and the interaction forces $\hat{\mathbf{p}}^{\mathbf{G}}$ given by [18]

$$\hat{\mathbf{p}}^{\mathbf{G}} = p^{GR} \text{grad } n^{\mathbf{G}} - \alpha_{\mathbf{wGS}} \mathbf{wGS} \tag{4.93}$$

where $\alpha_{\mathbf{wGS}}$ is a positive material parameter associated with the relative permeability of the gas phase are substituted into Equation (4.36) to yield

$$-\text{div}(n^{\mathbf{G}} p^{GR} \mathbf{I}) + \rho^{\mathbf{G}} \mathbf{b} + p^{GR} \text{grad } n^{\mathbf{G}} - \alpha_{\mathbf{wGS}} \mathbf{wGS} = 0. \tag{4.94}$$

Using the substitution

$$\text{div}(n^{\mathbf{G}} p^{GR} \mathbf{I}) = p^{GR} \text{grad } n^{\mathbf{G}} + n^{\mathbf{G}} \text{grad } p^{GR},$$

Equation (4.94) becomes

$$-p^{GR} \text{grad } n^{\mathbf{G}} - n^{\mathbf{G}} \text{grad } p^{GR} + \rho^{\mathbf{G}} \mathbf{b} + p^{GR} \text{grad } n^{\mathbf{G}} - \alpha_{\mathbf{wGS}} \mathbf{wGS} = 0. \tag{4.95}$$

The equation is rearranged to obtain the seepage velocity of the gas phase $n^{\mathbf{G}}\mathbf{w_{GS}}$ as

$$n^{\mathbf{G}}\mathbf{w_{GS}} = \frac{n^{\mathbf{G}}}{\alpha_{\mathbf{w_{GS}}}}[-n^{\mathbf{G}}\mathrm{grad}p^{GR} + \rho^{\mathbf{G}}\mathbf{b}]. \tag{4.96}$$

Substituting for the gas density $\rho^{\mathbf{G}} = n^{\mathbf{G}}\rho^{GR}$ gives

$$n^{\mathbf{G}}\mathbf{w_{GS}} = \frac{(n^{\mathbf{G}})^2}{\alpha_{\mathbf{w_{GS}}}}[-\mathrm{grad}p^{GR} + \rho^{GR}\,\mathbf{b}]. \tag{4.97}$$

Comparison of Equation (4.97) with *Darcy's* Law in Equation 2.1 leads to the ansatz

$$\frac{(n^{\mathbf{G}})^2}{\alpha_{\mathbf{w_{GS}}}} = \frac{\mathbf{k}k^{r\mathbf{G}}}{\mu^{\mathbf{G}}}, \tag{4.98}$$

which we refer to as the gaseous *Darcy* coefficient. In this equation, $\mathbf{k}$ is the intrinsic permeability also referred to as the absolute permeability of the porous solid body, $k^{r\mathbf{G}}$ and μ are respectively the relative permeability and the viscosity of the gas phase. The seepage velocity for the gas finally becomes

$$\boxed{n^{\mathbf{G}}\mathbf{w_{GS}} = \frac{\mathbf{k}k^{r\mathbf{G}}}{\mu^{\mathbf{G}}}[-n^{\mathbf{G}}\mathrm{grad}p^{GR} + \rho^{\mathbf{G}}\mathbf{b}].} \tag{4.99}$$

Similarly to what was done for the gas phase, the seepage velocity for the liquid phase is obtained from the balance of momentum for the solvent in Equation (4.34). The interaction forces $\hat{\mathbf{p}}^L$ are given by [18]

$$\hat{\mathbf{p}}^L = p^{LR}\mathrm{grad}\,n^L - \mu^{LQ}\mathrm{grad}\,n^L + \hat{\mathbf{p}}_E^L \tag{4.100}$$

$$\text{with} \quad \hat{\mathbf{p}}_E^L = -\gamma_{\bigtriangledown\theta}^L\,\mathrm{grad}\,\theta - \gamma_{\mathbf{w}_{LS}}^L\mathbf{w}_{LS} + \gamma_{\mathbf{w}_{LQS}}^L\mathbf{w}_{LQS} \tag{4.101}$$

where $\hat{\mathbf{p}}_E^L$ is the effective production of momentum for the solvent, $\gamma_{\bigtriangledown\theta}^L$ is the thermal conductivity of the solvent and $\gamma_{\mathbf{w}_{LS}}^L$ and $\gamma_{\mathbf{w}_{LQS}}^L$ are weighting factors for the influence of each velocity $\mathbf{w}_{LS}$ and $\mathbf{w}_{LQS}$ on the interaction forces. $\gamma_{\mathbf{w}_{LS}}^L$ and $\gamma_{\mathbf{w}_{LQS}}^L$ are always restricted to be greater than zero. Substituting the above and Equation (4.91) into Equation (4.34) gives

$$\begin{aligned}
&\mathrm{div}[-p^{LR}n^{\mathbf{L}}\mathbf{I} + \mu^{LQ}n^{\mathbf{L}}\mathbf{I}] + \rho^L\mathbf{b} + p^{LR}\mathrm{grad}\,n^L - \mu^{LQ}\mathrm{grad}\,n^L - \gamma_{\bigtriangledown\theta}^L\mathrm{grad}\,\theta \\
&- \gamma_{\mathbf{w}_{LS}}^L\mathbf{w}_{LS} + \gamma_{\mathbf{w}_{LQS}}^L\mathbf{w}_{LQS} = 0.
\end{aligned} \tag{4.102}$$

Evaluating

$$\mathrm{div}[-p^{LR}n^{\mathbf{L}}\mathbf{I} + \mu^{LQ}n^{\mathbf{L}}\mathbf{I}] = (\mu^{LQ}n^{\mathbf{L}}\delta_{ij})_{,j} - (p^{LR}n^{\mathbf{L}}\delta_{ij})_{,j}$$
$$= n^{\mathbf{L}}\mu^{LQ}_{,j}\delta_{ij} + \mu^{LQ}n^{\mathbf{L}}_{,j}\delta_{ij} - n^{\mathbf{L}}p^{LR}_{,j}\delta_{ij} - p^{LR}n^{\mathbf{L}}_{,j}\delta_{ij} \qquad (4.103)$$
$$= n^{\mathbf{L}}\mathrm{grad}\,\mu^{LQ} + \mu^{LQ}\mathrm{grad}\,n^{\mathbf{L}} - n^{\mathbf{L}}\mathrm{grad}\,p^{LR} - p^{LR}\mathrm{grad}\,n^{\mathbf{L}}$$

and substituting into Equation (4.102) gives;

$$n^{\mathbf{L}}\mathrm{grad}\,\mu^{LQ} + \mu^{LQ}\mathrm{grad}\,n^{\mathbf{L}} - n^{\mathbf{L}}\mathrm{grad}\,p^{LR} - p^{LR}\mathrm{grad}\,n^{\mathbf{L}} + \rho^{L}\mathbf{b} + p^{LR}\mathrm{grad}\,n^{\mathbf{L}} - \mu^{LQ}\mathrm{grad}\,n^{L}$$
$$- \gamma^{L}_{\nabla\theta}\mathrm{grad}\theta - \gamma^{L}_{\mathbf{w}_{LS}}\mathbf{w}_{LS} + \gamma^{L}_{\mathbf{w}_{LQS}}\mathbf{w}_{LQS} = 0.$$

$$(4.104)$$

For an isothermal process grad $\theta = 0$, hence Equation (4.104) is simplified to yield

$$n^{\mathbf{L}}\mathrm{grad}\,\mu^{LQ} - n^{\mathbf{L}}\mathrm{grad}\,p^{LR} + \rho^{L}\mathbf{b} - \gamma^{L}_{\mathbf{w}_{LS}}\mathbf{w}_{LS} + \gamma^{L}_{\mathbf{w}_{LQS}}\mathbf{w}_{LQS} = 0. \qquad (4.105)$$

To obtain the liquid seepage velocity $n^{\mathbf{L}}\mathbf{w}_{LS}$, Equation (4.105) is rearranged to obtain

$$n^{\mathbf{L}}\mathbf{w}_{LS} = \frac{n^{\mathbf{L}}}{\gamma^{L}_{\mathbf{w}_{LS}}}[n^{\mathbf{L}}(\mathrm{grad}\,\mu^{LQ} - \mathrm{grad}\,p^{LR}) + n^{\mathbf{L}}\rho^{LR}\mathbf{b} + \gamma^{L}_{\mathbf{w}_{LQS}}\mathbf{w}_{LQS}]. \qquad (4.106)$$

Comparison of the above equation with *Darcy's* Law in Equation (2.1) similarly to what was done with the gas phase yields

$$\frac{n^{\mathbf{L}}}{\gamma^{L}_{\mathbf{w}_{LS}}} = \frac{\mathbf{k}k^{rL}}{\mu^{L}}, \qquad (4.107)$$

which is referred to as the liquid *Darcy* coefficient. The parameters k^{rL} and μ^{L} are the relative permeability and the viscosity of the liquid phase respectively. The final form of the equation for the liquid seepage velocity $n^{\mathbf{L}}\mathbf{w}_{LS}$ is therefore

$$\boxed{n^{\mathbf{L}}\mathbf{w}_{LS} = \frac{\mathbf{k}k^{rL}}{\mu^{L}}[n^{\mathbf{L}}(\mathrm{grad}\,\mu^{LQ} - \mathrm{grad}\,p^{LR}) + n^{\mathbf{L}}\rho^{LR}\mathbf{b} + \gamma^{L}_{\mathbf{w}_{LQS}}\mathbf{w}_{LQS}].} \qquad (4.108)$$

Following similar steps to the solvent seepage velocity, the seepage velocity for the solute is obtained using Equation (4.35) with the interaction forces $\hat{\mathbf{p}}^{LQ}$ given by [18]

$$\hat{\mathbf{p}}^{LQ} = \mu^{LQ}\mathrm{grad}\,n^{\mathbf{L}} + \hat{\mathbf{p}}_E^{LQ} \tag{4.109}$$

$$\text{and}\quad \hat{\mathbf{p}}_E^{LQ} = -\gamma_{\nabla\theta}^Q\mathrm{grad}\theta - \gamma_{\mathbf{w}_{LS}}^Q\mathbf{w}_{LS} - \gamma_{\mathbf{w}_{LQS}}^Q\mathbf{w}_{LQS}. \tag{4.110}$$

Substituting Equations (4.90), (4.109) and (4.110) into Equation (4.35) yields

$$\mathrm{div}[-n^{\mathbf{L}}\mu^{LQ}\mathbf{I}] + \mu^{LQ}\mathrm{grad}\,n^{\mathbf{L}} - \gamma_{\nabla\theta}^Q\mathrm{grad}\theta - \gamma_{\mathbf{w}_{LS}}^Q\mathbf{w}_{LS} - \gamma_{\mathbf{w}_{LQS}}^Q\mathbf{w}_{LQS} = 0. \tag{4.111}$$

This is reformulated to give

$$- \mu^{LQ}\mathrm{grad}\,n^{\mathbf{L}} - n^{\mathbf{L}}\mathrm{grad}\,\mu^{LQ} + \mu^{LQ}\mathrm{grad}\,n^{\mathbf{L}} - \gamma_{\nabla\theta}^Q\mathrm{grad}\theta - \gamma_{\mathbf{w}_{LS}}^Q\mathbf{w}_{LS}$$
$$-\gamma_{\mathbf{w}_{LQS}}^Q\mathbf{w}_{LQS} = 0. \tag{4.112}$$

Recalling from the seepage velocity of the liquid that grad $\theta = 0$ gives the seepage velocity for the solute $n^{\mathbf{L}}\mathbf{w}_{LQS}$ as

$$\boxed{n^{\mathbf{L}}\mathbf{w}_{LQS} = \frac{n^{\mathbf{L}}}{\gamma_{\mathbf{w}_{LQS}}^Q}\left[-n^{\mathbf{L}}\mathrm{grad}\,\mu^{LQ} - \gamma_{\mathbf{w}_{LS}}^Q\mathbf{w}_{LS}\right].} \tag{4.113}$$

4.3.4 Liquid saturation

The choice of capillary pressure as the moisture state variable was discussed in Section 2.4.1. Ehlers and Bluhm [38] defined the capillary pressure p^C as

$$p^C = p^{\mathbf{GR}} - p^{LR}, \tag{4.114}$$

and couples the effective liquid to the effective gas pressures. The capillary pressure is further related to the degree of saturation with the liquid $s^{\mathbf{L}}$ which quantifies the proportion of fluids present in the pores using the *van Genuchten* relation

$$\boxed{s^{\mathbf{L}} = [1 + (\alpha p^C)^j]^{-h}} \tag{4.115}$$

with j and h being coupled according to the equation

$$h = 1 - \frac{1}{j} \tag{4.116}$$

where α and j are material constants determined experimentally.

4.4 Weak formulations of the governing equations

Considering the assumptions and constitutive relations set out in the previous sections, the unknown quantities are

$$\mathcal{U}(\mathbf{x}, t) = \{\mathbf{u_S}, \mathbf{w}_{LS}, \mathbf{w}_{LQS}, \mathbf{w}_{GS}, p^C, p^{GR}, n^{\mathbf{S}}, n^{\mathbf{L}}, n^{\mathbf{G}}, \mu^{LQ}\}. \tag{4.117}$$

The incompressibility of the solid phase enables the derivation of the solid volume fraction $n^{\mathbf{S}}$ in the spatial configuration by integrating the balance of mass for the solid phase to yield [16]

$$n^{\mathbf{S}} = \frac{n^{\mathbf{S}}_{0\mathbf{S}}}{J_{\mathbf{S}}}. \tag{4.118}$$

The liquid volume fraction $n^{\mathbf{L}}$ is calculated from the pore volume fraction $n^{\mathbf{F}} = 1 - n^{\mathbf{S}}$ and the degree of saturation $s^{\mathbf{L}}$ (obtained from the *van Genuchten* relation) as

$$n^{\mathbf{L}} = n^{\mathbf{F}} s^{\mathbf{L}}. \tag{4.119}$$

The gas volume fraction may then be obtained from the saturation condition according to;

$$n^{\mathbf{G}} = 1 - n^{\mathbf{S}} - n^{\mathbf{L}} \tag{4.120}$$

or similarly to the liquid volume fraction as

$$n^{\mathbf{G}} = n^{\mathbf{F}}(1 - s^{\mathbf{L}}). \tag{4.121}$$

The filter velocities $\mathbf{w}_{LS}$, $\mathbf{w}_{LQS}$ and $\mathbf{w}_{GS}$ are evaluated as described in Section 4.3.3. Therefore the remaining unknowns are

$$\mathcal{U}(\mathbf{x}, t) = \{\mathbf{u_S}, p^C, p^{GR}, \mu^{LQ}\}. \tag{4.122}$$

In order to solve for the variables in Equation (4.122), weak formulations based on the standard *Galerkin* procedure are formulated. The balance of momentum for the mixture is multiplied with the test function $\delta \mathbf{u_S}$ to solve for the solid displacements, $\mathbf{u_S}$. The volume balance for the liquid solvent is multiplied with the test function δp^C to solve for the capillary pressure, p^C and the mass balance for the pore gas is multiplied with the test function $\delta p^{\mathbf{GR}}$ to solve for the effective gas pressure, $p^{\mathbf{GR}}$. Finally, the balance of mass for the solute is multiplied with the test function $\delta \mu^{LQ}$ to solve for the chemical potential due to the concentration of the solute, μ^{LQ}.

4.4.1 Weak form for the balance of momentum for the mixture

The balance of linear momentum for the mixture is given in Equation (4.30). The weak form of Equation (4.30) is introduced as

$$G_{\mathbf{u_S}} = \int_B (\text{div } \mathbf{T} + \rho \mathbf{b}) \cdot \delta \mathbf{u_S} \, dv. \tag{4.123}$$

Proceeding to map to the reference configuration

$$\int_B \text{div } \mathbf{T} \, dv = \int_{\partial B} \mathbf{Tn} \, da = \int_{\partial B} \mathbf{Tn} \, J_{\mathbf{S}} \, \mathbf{F}^{T-1} dA_{0\mathbf{S}} = \int_{\partial B} \mathbf{PN} \, dA_{0\mathbf{S}} - \int_{\partial B} \text{Div } \mathbf{P} \, dV_{0\mathbf{S}} \tag{4.124}$$

where $\mathbf{n}$ and $\mathbf{N}$ are the outward surface normal vectors to the porous solid in the current and reference configuration respectively. Using the relationship

$$\text{Div } \mathbf{P} \cdot \delta \mathbf{u_S} = P_{ij,j} \, \delta u_i = (P_{ij} \, \delta u_i)_j - P_{ij} \, \delta u_{i,j} \tag{4.125}$$

$$\text{and} \quad \mathbf{P} \, \mathbf{N} = \mathbf{t} \tag{4.126}$$

where $\mathbf{t}$ is the traction [63] and *Green's* theorem

$$\int_{B_0} P_{ij} \delta u_{i,j} \, dV_{0S} = \int_{\partial B_0} P_{ij} \delta u_i \, N_j dA_{0S} = \int_{\partial B_0} t_i \delta u_i \, dA_{0S} \tag{4.127}$$

to obtain

$$\int_B \text{Div } \mathbf{P} \cdot \delta \mathbf{u_S} \, dv = \int_{B_0} \mathbf{P} \cdot \text{Grad } \delta \mathbf{u_S} \, dV_{0S} - \int_{\partial B_0} \mathbf{t} \cdot \delta \mathbf{u_S} \, dA_{0S} \tag{4.128}$$

yields

$$G_{\mathbf{us}} = \int_{B_0} \mathbf{P} \cdot \text{Grad } \delta \mathbf{us} \ dV_{0S} + \int_{B_0} J_S \, \rho \, \mathbf{b} \cdot \delta \mathbf{us} \ dV_{0S} - \int_{\partial B_0} \mathbf{t} \cdot \delta \mathbf{us} \ dA_{0S} = 0. \qquad (4.129)$$

4.4.2 Weak form for the balance of mass for the pore gas

The balance of mass of the gas phase is given in Equation (4.16). Taking its material time derivative with respect to the solid skeleton,

$$(\rho^{\mathbf{G}})'_{\mathbf{G}} = (\rho^{\mathbf{G}})'_{\mathbf{S}} + \text{grad } \rho^{\mathbf{G}} \cdot \mathbf{w_{GS}} \qquad (4.130)$$

and substituting this into Equation (4.16) yields

$$(\rho^{\mathbf{G}})'_{\mathbf{S}} + \text{grad } \rho^{\mathbf{G}} \cdot \mathbf{w_{GS}} + \rho^{\mathbf{G}} \text{ div } \mathbf{x}'_{\mathbf{G}} = 0. \qquad (4.131)$$

Recall that $\rho^{\mathbf{G}} = n^{\mathbf{G}} \rho^{GR}$ and that $n^{\mathbf{G}} = n^{\mathbf{F}} s^{\mathbf{G}}$ so that Equation (4.131) becomes

$$(n^{\mathbf{F}} s^{\mathbf{G}} \rho^{GR})'_{\mathbf{S}} + \text{grad } (n^{\mathbf{F}} s^{\mathbf{G}} \rho^{GR}) \cdot \mathbf{w_{GS}} + (n^{\mathbf{F}} s^{\mathbf{G}} \rho^{GR}) \text{ div } \mathbf{x}'_{\mathbf{G}} = 0. \qquad (4.132)$$

Expanding the above equation gives

$$(n^{\mathbf{F}})'_{\mathbf{S}} \, s^{\mathbf{G}} \rho^{GR} + (s^{\mathbf{G}})'_{\mathbf{S}} \, n^{\mathbf{F}} \rho^{GR} + (\rho^{GR})'_{\mathbf{S}} \, n^{\mathbf{F}} s^{\mathbf{G}} + \text{grad } (n^{\mathbf{F}} s^{\mathbf{G}} \rho^{GR}) \cdot \mathbf{w_{GS}}$$
$$+ (n^{\mathbf{F}} s^{\mathbf{G}} \rho^{GR}) \text{ div } \mathbf{x}'_{\mathbf{G}} = 0. \qquad (4.133)$$

Using $\text{div } \mathbf{x}'_{\mathbf{G}} = \text{div } (\mathbf{x}'_{\mathbf{S}} + \mathbf{w_{GS}})$, Equation (4.133) becomes

$$(n^{\mathbf{F}})'_{\mathbf{S}} \, s^{\mathbf{G}} \rho^{GR} + (s^{\mathbf{G}})'_{\mathbf{S}} \, n^{\mathbf{F}} \rho^{GR} + (\rho^{GR})'_{\mathbf{S}} \, n^{\mathbf{F}} s^{\mathbf{G}} + \text{grad}(n^{\mathbf{F}} s^{\mathbf{G}} \rho^{GR}) \cdot \mathbf{w_{GS}}$$
$$+ (n^{\mathbf{F}} s^{\mathbf{G}} \rho^{GR}) \text{ div } (\mathbf{x}'_{\mathbf{S}} + \mathbf{w_{GS}}) = 0. \qquad (4.134)$$

Equation (4.134) is divided by $s^{\mathbf{G}} \rho^{GR}$ and use is made of *Gauss*'s theorem to expand

$$\text{grad } (n^{\mathbf{F}} s^{\mathbf{G}} \rho^{GR}) \cdot \mathbf{w_{GS}} = \text{div } (n^{\mathbf{F}} s^{\mathbf{G}} \rho^{GR} \mathbf{w_{GS}}) - (n^{\mathbf{F}} s^{\mathbf{G}} \rho^{GR}) \text{ div } \mathbf{w_{GS}} \qquad (4.135)$$

and obtain

$$(n^{\mathbf{F}})'_{\mathbf{S}} + \frac{n^{\mathbf{F}}}{s^{\mathbf{G}}}(s^{\mathbf{G}})'_{\mathbf{S}} + \frac{n^{\mathbf{F}}}{\rho^{GR}}(\rho^{GR})'_{\mathbf{S}} + \frac{1}{s^{\mathbf{G}}\rho^{GR}}\mathrm{div}\,(n^{\mathbf{F}}s^{\mathbf{G}}\rho^{GR}\mathbf{w_{GS}}) - n^{\mathbf{F}}\,\mathrm{div}\,\mathbf{w_{GS}}$$
$$+\, n^{\mathbf{F}}\,\mathrm{div}\,(\mathbf{x}'_{\mathbf{S}} + \mathbf{w_{GS}}) = 0 \tag{4.136}$$

which may be rewritten as

$$(n^{\mathbf{F}})'_{\mathbf{S}} + \frac{n^{\mathbf{F}}}{s^{\mathbf{G}}}(s^{\mathbf{G}})'_{\mathbf{S}} + \frac{n^{\mathbf{F}}}{\rho^{GR}}(\rho^{GR})'_{\mathbf{S}} + \frac{1}{s^{\mathbf{G}}\rho^{GR}}\,\mathrm{div}\,(n^{\mathbf{F}}s^{\mathbf{G}}\rho^{GR}\mathbf{w_{GS}}) + n^{\mathbf{F}}\,\mathrm{div}\,\mathbf{x}'_{\mathbf{S}} = 0. \tag{4.137}$$

The mass balance of the solid phase is given as

$$-(n^{\mathbf{F}})'_{\mathbf{S}} + \frac{1-n^{\mathbf{F}}}{\rho^{SR}}(\rho^{SR})'_{\mathbf{S}} + (1-n^{\mathbf{F}})\,\mathrm{div}\,\mathbf{x}'_{\mathbf{S}} = 0. \tag{4.138}$$

Summation of equations (4.137) and (4.138) to eliminate $(n^{\mathbf{F}})'_{\mathbf{S}}$ yields

$$\frac{n^{\mathbf{F}}}{s^{\mathbf{G}}}\,(s^{\mathbf{G}})'_{\mathbf{S}} + \frac{n^{\mathbf{F}}}{\rho^{GR}}\,(\rho^{GR})'_{\mathbf{S}} + \frac{1}{s^{\mathbf{G}}\rho^{GR}}\,\mathrm{div}\,(n^{\mathbf{F}}s^{\mathbf{G}}\rho^{GR}\mathbf{w_{GS}}) + n^{\mathbf{F}}\,\mathrm{div}\,\mathbf{x}'_{\mathbf{S}}$$
$$\frac{1-n^{\mathbf{F}}}{\rho^{SR}}\,(\rho^{SR})'_{\mathbf{S}} + (1-n^{\mathbf{F}})\,\mathrm{div}\,\mathbf{x}'_{\mathbf{S}} = 0 \tag{4.139}$$

and is rewritten as

$$\frac{n^{\mathbf{F}}}{s^{\mathbf{G}}}(s^{\mathbf{G}})'_{\mathbf{S}} + \frac{n^{\mathbf{F}}}{\rho^{GR}}(\rho^{GR})'_{\mathbf{S}} + \frac{1}{s^{\mathbf{G}}\rho^{GR}}\mathrm{div}\,(n^{\mathbf{F}}s^{\mathbf{G}}\rho^{GR}\mathbf{w_{GS}}) + \frac{1-n^{\mathbf{F}}}{\rho^{SR}}(\rho^{SR})'_{\mathbf{S}} + \mathrm{div}\,\mathbf{x}'_{\mathbf{S}} = 0. \tag{4.140}$$

Recall from the constitutive theory that

$$\rho^{GR} = \frac{p^{GR}}{R\theta}$$

so that

$$(\rho^{GR})'_{\mathbf{S}} = \frac{1}{R\theta}(p^{GR})'_{\mathbf{S}}$$

and note that the material time derivative of the incompressible solid phase density is given by Lewis and Schrefler [37] as

$$\frac{1}{\rho^{SR}}(\rho^{SR})'_{\mathbf{S}} = \frac{1}{1-n^{\mathbf{F}}}\,\beta_{\mathbf{S}}\,(1-n^{\mathbf{F}})\,T'_{\mathbf{S}} \tag{4.141}$$

where T is the temperature above a reference value and β_S is the thermal coefficient of expansion for the solid phase [37].

Equation (4.141) is substituted into Equation (4.140) to obtain

$$\frac{n^{\mathbf{F}}}{s^{\mathbf{G}}}(s^{\mathbf{G}})'_{\mathbf{S}} + \frac{n^{\mathbf{F}}}{R\theta}(p^{GR})'_{\mathbf{S}} + \frac{1}{s^{\mathbf{G}}\rho^{GR}}\mathrm{div}\,(n^{\mathbf{F}}s^{\mathbf{G}}\rho^{GR}\mathbf{w_{GS}}) + \beta_{\mathbf{S}}(1-n^{\mathbf{F}})T'_{\mathbf{S}} + \mathrm{div}\,\mathbf{x'_S} = 0. \quad (4.142)$$

For an isothermal process, $T'_{\mathbf{S}} = 0$ so that the above equation becomes

$$\frac{n^{\mathbf{F}}}{s^{\mathbf{G}}}(s^{\mathbf{G}})'_{\mathbf{S}} + \frac{n^{\mathbf{F}}}{R\theta}(p^{GR})'_{\mathbf{S}} + \frac{1}{s^{\mathbf{G}}\rho^{GR}}\mathrm{div}\,(n^{\mathbf{F}}s^{\mathbf{G}}\rho^{GR}\mathbf{w_{GS}}) + \mathrm{div}\,\mathbf{x'_S} = 0 \quad (4.143)$$

Equation (4.143) is multiplied by $s^{\mathbf{G}}$ to obtain the weak form of the balance of mass of the gas phase as

$$G_{pGR} = \int_{B}[\,n^{\mathbf{F}}(s^{\mathbf{G}})'_{\mathbf{S}} + \frac{n^{\mathbf{F}}s^{\mathbf{G}}}{R\theta}(p^{GR})'_{\mathbf{S}} + \frac{1}{\rho^{GR}}\mathrm{div}\,(n^{\mathbf{F}}s^{\mathbf{G}}\rho^{GR}\mathbf{w_{GS}}) + s^{\mathbf{G}}\,\mathrm{div}\,\mathbf{x'_S}\,]\cdot\delta p^{GR}dv = 0 \quad (4.144)$$

Mapping to the reference configuration

$$\begin{aligned}
\int_{B}\mathrm{div}\,(n^{\mathbf{F}}s^{G}\rho^{GR}\mathbf{w_{GS}})\cdot\delta p^{\mathbf{GR}}dv &= \int_{\partial B}n^{\mathbf{F}}s^{G}\rho^{GR}\mathbf{w_{GS}}\cdot\mathbf{n}\,\delta p^{\mathbf{GR}}da \\
&= \int_{\partial B_0}n^{\mathbf{F}}s^{G}\rho^{GR}\mathbf{w_{GS0}}\cdot\mathbf{N}\,\delta p^{\mathbf{GR}}dA = \int_{B_0}\mathrm{Div}\,(n^{\mathbf{F}}s^{G}\rho^{GR}\mathbf{w_{GS0}})\cdot\delta p^{\mathbf{GR}}dV
\end{aligned} \quad (4.145)$$

with $n^{\mathbf{F}}s^{G}\mathbf{w_{GS}} = n^{\mathbf{F}}s^{G}\mathbf{w_{GS0}}NJ_{S}\mathbf{F}^{-T}$. Making use of *Gauss*'s theorem;

$$\mathrm{Div}(n^{\mathbf{F}}s^{G}\rho^{GR}\mathbf{w_{GS0}})\cdot\delta p^{\mathbf{GR}} = \mathrm{Div}(n^{\mathbf{F}}s^{G}\rho^{GR}\mathbf{w_{GS0}}\delta p^{GR}) - n^{\mathbf{F}}s^{G}\rho^{GR}\mathbf{w_{GS0}}\,\mathrm{Grad}\,\delta p^{\mathbf{GR}} \quad (4.146)$$

and

$$\int_{B_0}\mathrm{Div}(n^{\mathbf{F}}s^{G}\rho^{GR}\mathbf{w_{GS0}}\delta p^{GR})dV = \int_{\partial B_0}n^{\mathbf{F}}s^{G}\rho^{GR}\mathbf{w_{GS0}}\cdot\mathbf{N}\,\delta p^{\mathbf{GR}}dA. \quad (4.147)$$

Using $\mathrm{div}\,\mathbf{x'_S} = \mathrm{tr}\mathbf{D_S}$, the weak form becomes

$$\begin{aligned}
G_{pGR} = &\int_{B_0}[\,n^{\mathbf{F}}(s^{\mathbf{G}})'_{\mathbf{S}} + \frac{n^{\mathbf{F}}s^{\mathbf{G}}}{R\theta}(p^{GR})'_{\mathbf{S}}]\,\delta p^{\mathbf{GR}}dV + \int_{\partial B_0}n^{\mathbf{F}}s^{G}\mathbf{w_{GS0}}\cdot\mathbf{N}\,\delta p^{\mathbf{GR}}dA \\
&- \int_{B_0}n^{\mathbf{F}}s^{G}\rho^{GR}\mathbf{w_{GS0}}\,\mathrm{Grad}\,\delta p^{\mathbf{GR}}dV + \int_{B_0}s^{G}J_{\mathbf{S}}\,\mathrm{tr}\mathbf{D_S}\,\delta p^{\mathbf{GR}}dV = 0.
\end{aligned} \quad (4.148)$$

Substituting for the seepage velocity

$$n^{\mathbf{G}}\mathbf{w_{GS}} = \frac{(n^{\mathbf{G}})^2}{\alpha_{\mathbf{wGS}}}[-\mathrm{grad}\ p^{GR} + \rho^{GR}\mathbf{b}]$$
(4.149)

into Equation (4.148) yields;

$$G_{p^{GR}} = \int_{B_0}[\ n^{\mathbf{F}}(s^{\mathbf{G}})'_{\mathbf{S}} + \frac{n^{\mathbf{F}}s^{\mathbf{G}}}{R\theta}(p^{GR})'_{\mathbf{S}}]\ \delta p^{\mathbf{GR}}dV + \int_{\partial B_0} n^{\mathbf{F}}s^G\mathbf{w_{GS0}} \cdot \mathbf{N}\ \delta p^{\mathbf{GR}}dA$$
$$+ \int_{B_0}\frac{(n^{\mathbf{G}})^2}{\alpha_{\mathbf{wGS}}}\ \mathrm{Grad}\ p^{GR}\mathrm{Grad}\ \delta p^{GR}dV - \int_{B_0}\frac{(n^{\mathbf{G}})^2}{\alpha_{\mathbf{wGS}}}\rho^{GR}\mathbf{b}\ \mathrm{Grad}\ \delta p^{GR}dV$$
(4.150)
$$+ \int_{B_0} s^G\ J_{\mathbf{S}}\ \mathrm{tr}\mathbf{D_S}\ \delta p^{\mathbf{GR}}dV = 0$$

which is rewritten as

$$\int_{B_0}[\ n^{\mathbf{F}}(s^{\mathbf{G}})'_{\mathbf{S}} + \frac{n^{\mathbf{F}}s^{\mathbf{G}}}{R\theta}(p^{GR})'_{\mathbf{S}}]\ \delta p^{\mathbf{GR}}dV + \int_{B_0} s^G\ J_{\mathbf{S}}\ \mathrm{tr}\mathbf{D_S}\ \delta p^{\mathbf{GR}}dV$$
$$+ \int_{B_0}\frac{(n^{\mathbf{G}})^2}{\alpha_{\mathbf{wGS}}}\ \mathrm{Grad}\ p^{\mathbf{GR}}\mathrm{Grad}\ \delta p^{\mathbf{GR}}dV = \int_{B_0}\frac{(n^{\mathbf{G}})^2}{\alpha_{\mathbf{wGS}}}\rho^{GR}\mathbf{b}\ \mathrm{Grad}\ \delta p^{\mathbf{GR}}dV$$
(4.151)
$$- \int_{\partial B_0} n^{\mathbf{F}}s^G\mathbf{w_{GS0}} \cdot \mathbf{N}\ \delta p^{\mathbf{GR}}dA.$$

From Equation (4.5) and Lewis and Schrefler [37], $(s^{\mathbf{G}})'_{\mathbf{S}} = -(s^{\mathbf{L}})'_{\mathbf{S}}$ and

$$(s^{\mathbf{L}})'_{\mathbf{S}} = \frac{\partial s^{\mathbf{L}}}{\partial p^C}\frac{\partial p^C}{\partial t} = C_s(\frac{\partial p^{GR}}{\partial t} - \frac{\partial p^{LR}}{\partial t})$$
(4.152)

where C_s is the specific moisture content. Hence the final form of Equation (4.151) is

$$\boxed{\begin{aligned} G_{p^{\mathbf{G}}} &= \int_{B_0}[\ -n^{\mathbf{F}}C_s(\frac{\partial p^{GR}}{\partial t} - \frac{\partial p^{LR}}{\partial t}) + \frac{n^{\mathbf{F}}s^{\mathbf{G}}}{R\theta}(p^{GR})'_{\mathbf{S}}]\ \delta p^{\mathbf{GR}}dV + \int_{B_0} s^G\ J_{\mathbf{S}}\ \mathrm{tr}\mathbf{D_S}\ \delta p^{\mathbf{GR}}dV \\ &+ \int_{B_0}\frac{(n^{\mathbf{G}})^2}{\alpha_{\mathbf{wGS}}}\ \mathrm{Grad}\ p^{\mathbf{GR}}\mathrm{Grad}\ \delta p^{\mathbf{GR}}dV = \int_{B_0}\frac{(n^{\mathbf{G}})^2}{\alpha_{\mathbf{wGS}}}\rho^{GR}\mathbf{b}\ \mathrm{Grad}\ \delta p^{GR}dV \\ &- \int_{\partial B_0} n^{\mathbf{F}}s^G\mathbf{w_{GS0}} \cdot \mathbf{N}\ \delta p^{\mathbf{GR}}dA \end{aligned}}$$
(4.153)

4.4.3 Weak form for the balance of mass for the pore liquid

The balance of mass for the liquid phase is given in Equation (4.19). Taking its material time derivative with respect to the solid skeleton,

$$(\rho^L)'_L = (\rho^L)'_S + \mathrm{grad}\rho^L \cdot \mathbf{w}_{LS} \tag{4.154}$$

and substituting this into Equation (4.19) gives

$$(\rho^L)'_S + \mathrm{grad}\rho^L \cdot \mathbf{w}_{LS} + \rho^L \, \mathrm{div} \, \mathbf{x}'_L = 0. \tag{4.155}$$

Recall that $\rho^L = n^\mathbf{L}\rho^{LR}$ and $n^\mathbf{L} = n^\mathbf{F}s^\mathbf{L}$ so that Equation (4.155) becomes

$$(n^\mathbf{F}s^\mathbf{L}\rho^{LR})'_S + \mathrm{grad}(n^\mathbf{F}s^\mathbf{L}\rho^{LR}) \cdot \mathbf{w}_{LS} + (n^\mathbf{F}s^\mathbf{L}\rho^{LR}) \, \mathrm{div} \, \mathbf{x}'_L = 0 \tag{4.156}$$

which is rewritten as

$$(n^\mathbf{F})'_S s^\mathbf{L}\rho^{LR}+(s^\mathbf{L})'_S n^\mathbf{F}\rho^{LR}+(\rho^{LR})'_S \, s^\mathbf{L}n^\mathbf{F}+\mathrm{grad}(n^\mathbf{F}s^\mathbf{L}\rho^{LR})\cdot\mathbf{w}_{LS}+(n^\mathbf{F}s^\mathbf{L}\rho^{LR}) \, \mathrm{div} \, \mathbf{x}'_L = 0. \tag{4.157}$$

Dividing through by $s^\mathbf{L}\rho^{LR}$ and substituting for $\mathrm{div} \, \mathbf{x}'_L = \mathrm{div} \, (\mathbf{x}'_S + \mathbf{w}_{LS})$ results in

$$(n^\mathbf{F})'_S + \frac{n^\mathbf{F}}{s^\mathbf{L}}(s^\mathbf{L})'_S + \frac{n^\mathbf{F}}{\rho^{LR}}(\rho^{LR})'_S + \frac{1}{s^\mathbf{L}\rho^{LR}}\mathrm{grad}(n^\mathbf{F}s^\mathbf{L}\rho^{LR}) \cdot \mathbf{w}_{LS} + n^\mathbf{F} \, \mathrm{div} \, (\mathbf{x}'_S + \mathbf{w}_{LS}) = 0. \tag{4.158}$$

Summation of Equation (4.158) with Equation (4.138) to eliminate $(n^\mathbf{F})'_S$ yields

$$\begin{aligned}
&\frac{n^\mathbf{F}}{s^\mathbf{L}}(s^\mathbf{L})'_S + \frac{n^\mathbf{F}}{\rho^{LR}}(\rho^{LR})'_S + \frac{1}{s^\mathbf{L}\rho^{LR}}\mathrm{grad} \, (n^\mathbf{F}s^\mathbf{L}\rho^{LR}) \cdot \mathbf{w}_{LS} + n^\mathbf{F} \, \mathrm{div} \, (\mathbf{x}'_S + \mathbf{w}_{LS}) \\
&+ \frac{1-n^\mathbf{F}}{\rho^{SR}}(\rho^{SR})'_S + (1-n^\mathbf{F}) \, \mathrm{div} \, \mathbf{x}'_S = 0
\end{aligned} \tag{4.159}$$

Using

$$\mathrm{grad} \, (n^\mathbf{F}s^\mathbf{L}\rho^{LR}) \cdot \mathbf{w}_{LS} = \mathrm{div} \, (n^\mathbf{F}s^\mathbf{L}\rho^{LR}\mathbf{w}_{LS}) - n^\mathbf{F}s^\mathbf{L}\rho^{LR} \, \mathrm{div} \, \mathbf{w}_{LS},$$

Equation (4.159) is expressed in the form

$$\frac{n^{\mathbf{F}}}{s^{\mathbf{L}}}\,(s^{\mathbf{L}})'_{\mathbf{S}} + \frac{n^{\mathbf{F}}}{\rho^{LR}}\,(\rho^{LR})'_{\mathbf{S}} + \frac{1}{s^{\mathbf{L}}\rho^{LR}}\,\mathrm{div}\,(n^{\mathbf{F}}s^{\mathbf{L}}\rho^{LR}\mathbf{w}_{LS}) - n^{\mathbf{F}}\mathrm{div}\,\mathbf{w}_{LS} + n^{\mathbf{F}}\,\mathrm{div}\,\mathbf{w}_{LS}$$
$$+ \frac{1-n^{\mathbf{F}}}{\rho^{SR}}\,(\rho^{SR})'_{\mathbf{S}} + \mathrm{div}\,\mathbf{x}'_{\mathbf{S}} = 0 \tag{4.160}$$

and is simplified to become

$$\frac{n^{\mathbf{F}}}{s^{\mathbf{L}}}\,(s^{\mathbf{L}})'_{\mathbf{S}} + \frac{n^{\mathbf{F}}}{\rho^{LR}}\,(\rho^{LR})'_{\mathbf{S}} + \frac{1}{s^{\mathbf{L}}\rho^{LR}}\,\mathrm{div}\,(n^{\mathbf{F}}s^{\mathbf{L}}\rho^{LR}\mathbf{w}_{LS}) + \frac{1-n^{\mathbf{F}}}{\rho^{SR}}\,(\rho^{SR})'_{\mathbf{S}} + \mathrm{div}\,\mathbf{x}'_{\mathbf{S}} = 0. \tag{4.161}$$

The material time derivative of the incompressible solid density is given in Equation (4.141) and that of the incompressible liquid density by

$$\frac{1}{\rho^{LR}}(\rho^{LR})'_{\mathbf{S}} = \frac{1}{K_L}(p^{LR})'_{\mathbf{S}} - \beta_L T'_{\mathbf{S}} \tag{4.162}$$

where K_L is the bulk modulus of the liquid phase and β_L is its coefficient of thermal expansion [37]. Inserting these into Equation (4.161) yields

$$\frac{n^{\mathbf{F}}}{s^{\mathbf{L}}}(s^{\mathbf{L}})'_{\mathbf{S}} + \frac{n^{\mathbf{F}}}{K_L}(p^{LR})'_{\mathbf{S}} - \beta_L T'_{\mathbf{S}} + \frac{1}{s^{\mathbf{L}}\rho^{LR}}\mathrm{div}\,(n^{\mathbf{F}}s^{\mathbf{L}}\rho^{LR}\mathbf{w}_{LS}) + \beta_{\mathbf{S}}(1-n^{\mathbf{F}})T'_{\mathbf{S}} + \mathrm{div}\,\mathbf{x}'_{\mathbf{S}} = 0. \tag{4.163}$$

For an isothermal process, $T'_{\mathbf{S}} = 0$ so that Equation (4.163) becomes

$$\frac{n^{\mathbf{F}}}{s^{\mathbf{L}}}(s^{\mathbf{L}})'_{\mathbf{S}} + \frac{n^{\mathbf{F}}}{K_L}(p^{LR})'_{\mathbf{S}} + \frac{1}{s^{\mathbf{L}}\rho^{LR}}\mathrm{div}\,(n^{\mathbf{F}}s^{\mathbf{L}}\rho^{LR}\mathbf{w}_{LS}) + \mathrm{div}\,\mathbf{x}'_{\mathbf{S}} = 0. \tag{4.164}$$

Multiplying Equation (4.164) by $s^{\mathbf{L}}$ yields the weak formulation for the balance of mass of the pore liquid as

$$G_{p^C} = \int_B [n^{\mathbf{F}}(s^{\mathbf{L}})'_{\mathbf{S}} + \frac{n^{\mathbf{F}}s^{\mathbf{L}}}{K_L}(p^{LR})'_{\mathbf{S}} + \frac{1}{\rho^{LR}}\mathrm{div}\,(n^{\mathbf{F}}s^{\mathbf{L}}\rho^{LR}\mathbf{w}_{LS}) + s^{\mathbf{L}}\mathrm{div}\,\mathbf{x}'_{\mathbf{S}}]\cdot\delta p^C\;\mathrm{d}v = 0 \tag{4.165}$$

Mapping to the reference configuration and making use of *Gauss*'s theorem as was done for the weak form of the gas yields

$$G_{p^C} = \int_{B_0} [n^{\mathbf{F}}(s^{\mathbf{L}})'_{\mathbf{S}} + \frac{n^{\mathbf{F}}s^{\mathbf{L}}}{K_L}(p^{LR})'_{\mathbf{S}}]\cdot\delta p^C\;\mathrm{d}V + \int_{\partial B_0} n^{\mathbf{F}}s^{\mathbf{L}}\mathbf{w}_{LS0}\cdot\mathbf{N}\,\delta p^C\mathrm{d}A$$
$$- \int_{B_0} n^{\mathbf{F}}s^{\mathbf{L}}\mathbf{w}_{LS0}\mathrm{Grad}\,\delta p^C\mathrm{d}V + \int_{B_0} s^{\mathbf{L}}J_{\mathbf{S}}\mathrm{div}\,\mathbf{x}'_{\mathbf{S}}\,\delta p^C\;\mathrm{d}V = 0 \tag{4.166}$$

Substituting Equation (4.152) for the material time derivative of the liquid saturation degree in the above equation yields

$$
\begin{aligned}
G_{pC} = &\int_{B_0} [[n^{\mathbf{F}} C_s(\frac{\partial p^{GR}}{\partial t} - \frac{\partial p^{LR}}{\partial t}) + \frac{n^{\mathbf{F}} s^{\mathbf{L}}}{K_L}(p^{LR})'_{\mathbf{S}}] \cdot \delta p^C \, \mathrm{d}V + \int_{\partial B_0} n^{\mathbf{F}} s^{\mathbf{L}} \mathbf{w}_{LS0} \cdot \mathbf{N} \, \delta p^C \mathrm{d}A \\
&- \int_{B_0} n^{\mathbf{F}} s^{\mathbf{L}} \mathbf{w}_{LS0} \mathrm{Grad}\, \delta p^C \mathrm{d}V + \int_{B_0} s^{\mathbf{L}} J_{\mathbf{S}} \mathrm{div}\, \mathbf{x}'_{\mathbf{S}} \, \delta p^C \, \mathrm{d}V = 0
\end{aligned}
\tag{4.167}
$$

4.4.4 Weak form for the balance of mass for the solute

The mass balance for the solute is given in Equation (4.27) and its material time derivative with respect to the solid skeleton by

$$
(n^{\mathbf{L}} c^{LQ})'_{\mathbf{S}} + \mathrm{div}\, (n^{\mathbf{L}} c^{LQ} \mathbf{w}_{LQ\mathbf{S}}) + n^{\mathbf{L}} c^{LQ} \, \mathrm{div}\, \mathbf{x}'_{\mathbf{S}} - \hat{\rho}^{LQ}_{mol} = 0
\tag{4.168}
$$

with

$$
\hat{\rho}^{LQ}_{mol} = \frac{\hat{\rho}^{LQ}}{M^{Q}_{mol}} = 0
$$

and

$$
\mathbf{j}^{LQ} = n^{\mathbf{L}} c^{LQ} \mathbf{w}_{LQ\mathbf{S}}
\tag{4.169}
$$

as the diffusive flux. The weak form of the balance of mass for the solute is derived to be

$$
G_{\mu^{LQ}} = \int_{B} [(n^{\mathbf{L}} c^{LQ})'_{\mathbf{S}} + \mathrm{div}\, \mathbf{j}^{LQ} + n^{\mathbf{L}} c^{LQ} \, \mathrm{div}\, \mathbf{x}'_{\mathbf{S}}] \cdot \delta \mu^{LQ} \, \mathrm{d}v = 0
\tag{4.170}
$$

Mapping to the reference configuration

$$
\begin{aligned}
\int_{B} \mathrm{div}\, \mathbf{j}^{LQ} \, \mathrm{d}v &= \int_{\partial B} \mathrm{div}\, \mathbf{j}^{LQ} \cdot \mathbf{n}\, \mathrm{d}a = \int_{\partial B} \mathbf{j}^{LQ} \mathbf{N}\, J_{\mathbf{S}} \mathbf{F}^{T-1} \, \mathrm{d}A_{0S} = \int_{\partial B} \mathbf{j}^{LQ}_0 \mathbf{N}\, \mathrm{d}A_{0S} \\
&= \int_{B} \mathrm{Div}\, \mathbf{j}^{LQ}_0 \, \mathrm{d}V_{0S}
\end{aligned}
\tag{4.171}
$$

with

$$
\mathbf{j}^{LQ}_0 = \mathbf{j}^{LQ} \mathbf{N}\, J_{\mathbf{S}} \mathbf{F}^{T-1} = n^{\mathbf{L}} c^{LQ} \mathbf{w}_{LQ\mathbf{S}0}
\tag{4.172}
$$

Applying *Gauss'* theorem;

$$\text{Div } \mathbf{j}_0^{LQ} \cdot \delta\mu^{LQ} = \text{Div } \mathbf{j}_0^{LQ} \, \delta\mu^{LQ} - \mathbf{j}_0^{LQ} \cdot \text{Grad } \delta\mu^{LQ} \tag{4.173}$$

and

$$\int_{B_0} \text{Div } \mathbf{j}_0^{LQ} \, \delta\mu^{LQ} \, dV_{0S} = \int_{\partial B_0} \mathbf{j}_0^{LQ} \cdot \mathbf{N} \, \delta\mu^{LQ} \, dA_{0S} \tag{4.174}$$

along with div $\mathbf{x}'_{\mathbf{S}} = tr\mathbf{D_S}$ yields the weak form of the balance of mass for the solute as

$$\boxed{\begin{aligned} G_{\mu^{LQ}} = & -\int_{B_0} \mathbf{j}_0^{LQ} \cdot \text{Grad } \delta\mu^{LQ} \, dV_{0S} + \int_{B_0} n^{\mathbf{L}} c^{LQ} J_{\mathbf{S}} \, tr\mathbf{D_S}\delta\mu^{LQ} \, dV_{0S} \\ & + \int_{B_0} (n^{\mathbf{L}} c^{LQ})'_{\mathbf{S}} J_{\mathbf{S}} \, \delta\mu^{LQ} \, dV_{0S} + \int_{\partial B_0} \mathbf{j}_0^{LQ} \cdot \mathbf{N} \, \delta\mu^{LQ} \, dA_{0S} \end{aligned}} \tag{4.175}$$

5. Numerical treatment

5.1 Discretization

Implementation of the triphasic framework was performed using finite element code in the in-house software *SESKA*. The spatial discretization made use of quadratic test functions for the solid displacements and linear test functions for the capillary pressure and gas pressure yielding first-order partial differential equations.

The Newmark[1] method was chosen as the time integration scheme in the finite element approximation. It employs the extended mean value theorem to solve for displacements $\bar{\mathbf{d}}_{n+1}$ and velocity $\bar{\mathbf{v}}_{n+1}$ at time t_{n+1} from the equation of motion as

$$\bar{\mathbf{d}}_{n+1} = \tilde{\mathbf{d}}_{n+1} + \Delta t^2 \, \beta \, \bar{\mathbf{a}}_{n+1} \tag{5.1}$$

$$\bar{\mathbf{v}}_{n+1} = \tilde{\mathbf{v}}_{n+1} + \gamma \, \Delta t \, \bar{\mathbf{a}}_{n+1} \tag{5.2}$$

where Δt is the time increment with

$$\tilde{\mathbf{d}}_{n+1} = \bar{\mathbf{d}}_n + \Delta t \, \bar{\mathbf{v}}_n + \frac{\Delta t^2}{2} \, (1 - 2\beta) \, \bar{\mathbf{a}}_n \tag{5.3}$$

$$\tilde{\mathbf{v}}_{n+1} = \bar{\mathbf{v}}_n + \Delta t \, (1 - \gamma) \, \bar{\mathbf{a}}_n \tag{5.4}$$

where $\ddot{\mathbf{u}} = \bar{\mathbf{a}}_n$ is the acceleration, $\dot{\mathbf{u}} = \bar{\mathbf{v}}_n$ is the velocity and $\mathbf{u} = \bar{\mathbf{d}}_n$ is the displacement at time t_n. For unconditional stability, the Newmark parameters were chosen to be $\gamma = \frac{1}{2}$ and $\beta = \frac{1}{4}$ [19].

---

[1] The Newmark method developed by Nathan Newmark in 1959 is a numerical integration scheme used to solve differential equations typically in the modelling of dynamic systems.

5.2 Simulation

The triphasic model excluding the solute concentration was implemented in *SESKA*. An attempt was made to simulate the drainage of a column based on the leaking problem in Ehlers and Bluhm [38]. The example presented here is academic in nature. The results demonstrate how the capillary, gas and liquid pressures function in this preliminary model and demonstrate some of the challenges encountered in its implementation thus far.

The material parameters chosen for the model are listed in Table 5.1.

TABLE 5.1: The material parameters were obtained from Ehlers and Bluhm [38].

Parameter	Symbol	Value
Lame constants	μ^S	$5583\ kN/m^2$
	λ^S	$8375\ kN/m^2$
Solid density	$\rho^{\mathbf{S}R}$	$2600\ kg/m^3$
Gas density	$\rho_0^{\mathbf{G}R}$	$1.23\ kg/m^3$
Liquid density	ρ^{LR}	$1000\ kg/m^3$
Gas constants	R^G	$287.17\ J/(kgK)$
	θ	$283\ K$
	p_0	$10^5\ N/m^2$
Solid volume fraction	$n_{0\mathbf{S}}^{\mathbf{S}}$	0.67
Pore volume fraction	n_{0F}^{F}	0.33
Gravity	g	$10\ m/s^2$
Intrinsic permeability	$\mathbf{k}$	$1.2 \times 10^{14}\ m/s^2$
Liquid viscosity	μ^{LR}	$10^{-3}\ Ns/m^2$
Gas viscosity	μ^{GR}	$1.8 \times 10^{-5}\ Ns/m^2$
Permeability parameters	α	2×10^{-4}
	j	2.3
	h	1.5
	p	5.5

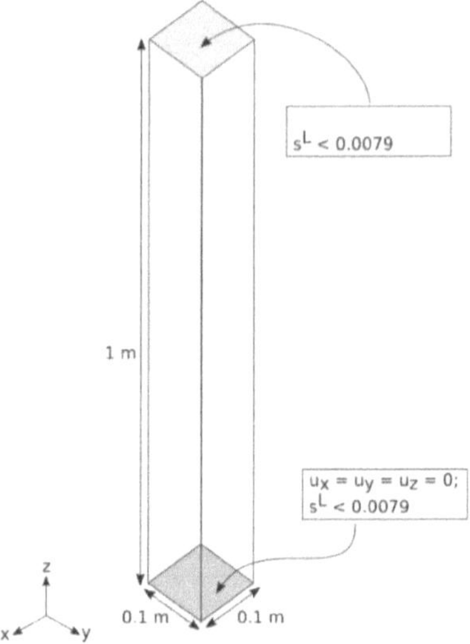

FIGURE 5.1: Problem definition showing the column dimensions and the boundary conditions applied.

Figure 5.1 shows the geometry and the boundary conditions applied in the simulation. The column had dimensions of 0.1 meter by 0.1 meter and a height of 1 meter. The bottom surface of the column was fixed in all the displacement degrees of freedom. The capillary pressure was set to $p^C = 20000MPa$ to induce a liquid saturation of $s^L = 0.0079$ on this surface and on the top surface of the column. The capillary pressure was ramped up to yield a decreasing liquid saturation towards $s^L = 0.0000$. Applying a saturation degree of $s^L = 0.0$ instantaneously produced numerical errors assumed to arise from the system becoming biphasic on these boundaries. No traction was applied. The solid saturation point was set at $s^L = 0.0$ as this was not specified in the reference.

The results of the simulation are presented in the next sections.

5.3 Results and discussion

Figure 5.2 shows the progression of the liquid saturation s^L during simulation of a leaking problem in Ehlers and Bluhm [38].

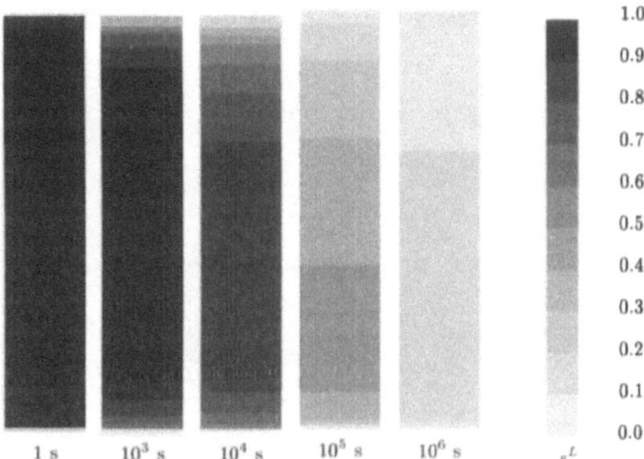

FIGURE 5.2: Development of liquid saturation for leaking problem in Ehlers and Bluhm [38]

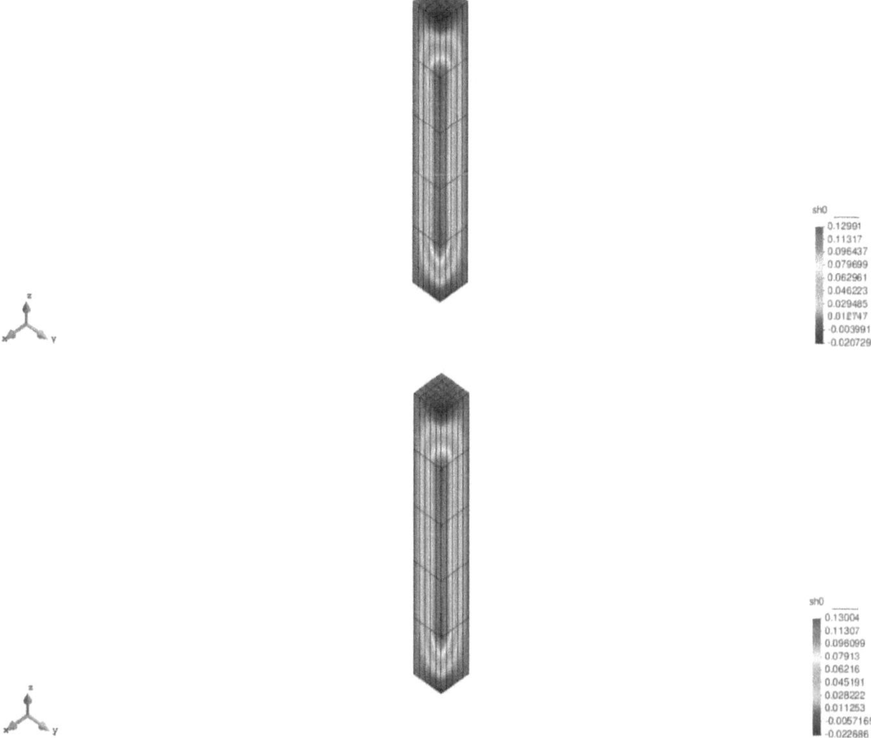

FIGURE 5.3: Liquid saturation at 0.001s and 0.59s

Figure 5.3 shows the liquid saturation (sh0 $= s^L$) at the beginning and at the termination of the simulation at 0.59s of the problem time. Figures 5.4 and 5.5 show the seepage velocities obtained by Ehlers and Bluhm [38] and our model respectively.

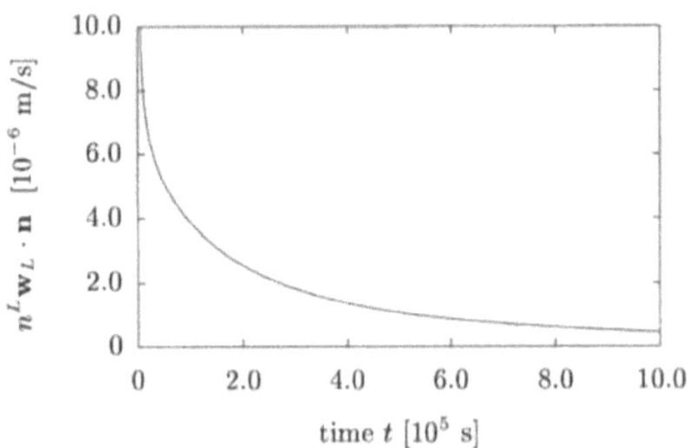

FIGURE 5.4: Seepage velocity vs. time obtained by Ehlers and Bluhm [38] at the bottom surface of the column

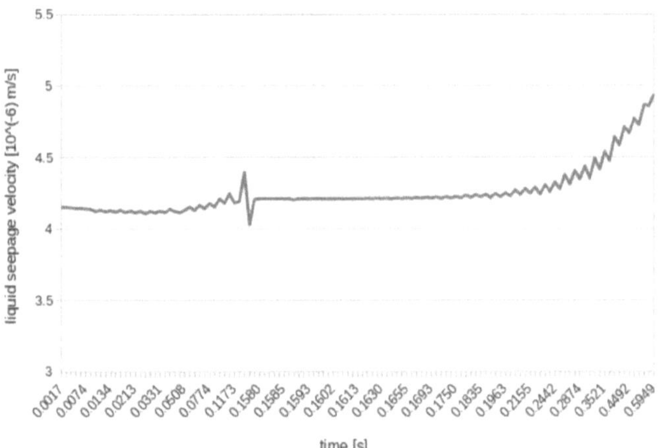

FIGURE 5.5: A graph showing the variation of the seepage velocity vs. time obtained from the simulation at the bottom surface of the column

A visual comparison between Figures 5.2, 5.3, 5.4 and 5.5 shows major discrepancies between the results obtained from the model and the expected results. Whereas a vertical gradation of the liquid saturation is observed in Figure 5.2, Figure 5.3 shows the liquid saturation to be radially decreasing from the center of the column outwards. This infers that the liquid is seeping out of

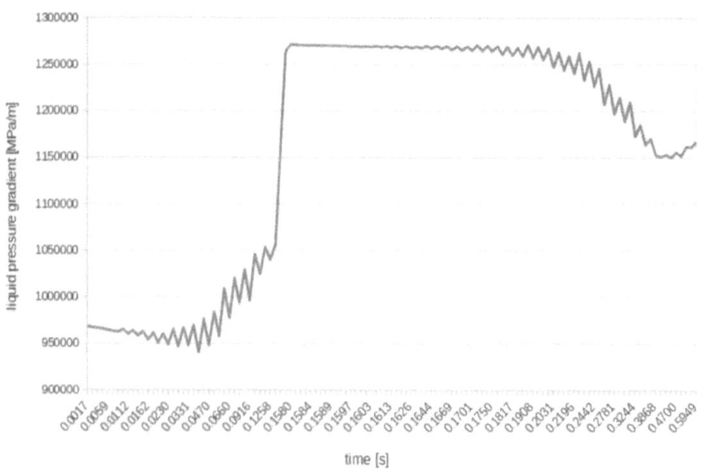

FIGURE 5.6: A graph showing the variation of the liquid pressure gradient vs. time obtained from the simulation at the bottom surface of the column

the column from all surfaces and is not the right behaviour. This may be due to time-scale or unit conversion errors made in the input file. It was expected that application of these boundary conditions as done in Ehlers and Bluhm [38] would yield a gravity drainage of the column resulting in the liquid saturation variation observed in Figure 5.2.

Comparison of Figures 5.4 and 5.5 shows that although the graph of the seepage velocity at the bottom surface of the column in Ehlers and Bluhm [38] was a smooth curve with a decreasing slope, the same graph obtained from the model is an oscillating curve with an increasing slope at the later stages of the simulation. The oscillations are a typical result of derived quantities in multiphase modelling and are usually smoothed by applying ramped up or down boundary conditions instead of constant boundary conditions. Although the capillary pressure boundary condition was ramped up, this was did not eliminate the oscillations as observed in Figure 5.5.

Referring to Equation (4.108) the seepage velocity of the pore liquid depends on the *Darcy* coefficient, the gradient of the liquid pressure and the body forces. No body forces were applied in this example and therefore the seepage velocity is dependent only on the *Darcy* coefficient and the liquid pressure gradient. The increase in the liquid seepage velocity is attributed to the ramped up capillary pressure boundary condition applied to this surface and the upper surface of the column. This boundary condition induced an increasing liquid pressure gradient and resulted in an increasing outflow of liquid.

Oscillations in the results of the liquid pressure gradient were observed and are shown in Figure 5.6. The liquid pressure gradient was calculated as the difference between the gas pressure gradient and the capillary pressure gradient. The gas pressure gradient showed oscillations identical to those of the liquid pressure gradient and was identified to be the cause of the oscillations in the liquid pressure gradient. These in turn produced oscillations in the liquid seepage velocity. From Figure 5.3, the gas is present in the column at saturations $s^G > 0.87$ $(s^G = 1 - s^L)$. The gas pressure gradient is a derived quantity and its oscillations were assumed to arise from the lack of gas pressure boundary conditions.

From the peculiar results presented in this work, it is evident that the model equations were not implemented correctly in *SESKA*. Due to the limited time-frame of the study, this matter could not be investigated further. Additional work is required on the implementation of the proposed model in *SESKA* to include the solute transport. An investigation into the influence of various boundary conditions and input parameters on the model behaviour needs to be conducted so that these can be accounted for in the verification and calibration of the model. It is recommended that a simpler example be used for verification of the model for the purpose of demonstrating the functionality of the model.

6. Conclusions and future work

6.1 Conclusions

The degradation of the protective oxide layer on reinforcing steel in concrete initiates corrosion. Corrosion products occupy a larger volume than the corroding steel and therefore create tensile stresses in the limited space of the steel-concrete interface. Studies have shown that not all the corrosion products contribute to these tensile stresses. A proportion of the corrosion products penetrates into the capillary pores in the porous zone of concrete adjacent to the steel. Although the mechanism by which the penetration occurs is not well understood, it has been shown that the major source of discrepancies between time-to-cracking service life models for reinforced concrete and experimental results is not sufficiently addressing this phenomena in the modelling effort.

Recent studies have endeavoured to account for the penetration of corrosion products in time-to-cracking concrete models. Each study reviewed utilised a different method for calculating the amount of corrosion products penetrating into the concrete pores and for calculating the size of the CAR. With so many varied approaches, both the amount of corrosion products penetrating into the concrete and the size of the CAR were often either underestimated or overestimated.

This work aimed to develop a theoretical framework for a numerical model to simulate the penetration of corrosion products into concrete pores using the Theory of Porous Media. Therefore a theoretical triphasic model with a miscible component based on TPM was derived. The three phases included in the model are the solid (concrete), liquid (pore solution) and gas (oxygen) phases. The soluble ion (iron III chloride ($FeCl_3^-$)) was added as a concentration in the liquid phase. The additional degrees of freedom namely the capillary pressure, the gas pressure and the chemical potential of the solute allowed coupling between the displacement fields and fluid flow.

An attempt was made to implement the triphasic TPM model excluding the solute in the in-house software *SESKA*. The leaking of a soil column was simulated as an academic example to demonstrate the functionality of the additional degrees of freedom in the model. The results obtained were inconsistent with the reference example. Due to the limited time-frame of this work, the inconsistencies between the reference example and the model results could not be addressed at this stage. The model presented and the results of the implementation cannot be conclusively validated until the numerical implementation is functioning correctly. This is the initial contribution towards the development of a multiphasic material model for the deterioration of reinforced concrete subjected to chloride-induced corrosion in *SESKA*.

6.2   Future work

The model presented needs to be refined, calibrated and validated for reinforced concrete. This may be done using experimental data and model results on the wetting and drying of reinforced concrete in which the relationship between capillary pressure and saturation is well documented. The most important parameters to note during calibration are the intrinsic permeability of the material and the relative permeability relationships as associated with the chosen sorption-desorption isotherm.

Following validation of the existing model, the solute phase will need to be implemented, calibrated and validated in *SESKA*. Although there is very little experimental data on the penetration of iron III chloride into the concrete pores, any ion may be used to demonstrate the functionality of the model. It is recommended that chloride ingress experimental results be used for this purpose instead, as the mechanisms by which both processes occur are similar. Alternatively, experimental tests may be performed in conjunction with the computational simulation.

The formulation presented within this work does not include mass exchange between the constituents of the porous medium. However, to realistically model the the stresses due to precipitation of iron III chloride within the concrete pore structure, mass exchange between the solute, liquid and solid phases will have to be introduced within the model.

To extend this model to a comprehensive time-to-cracking service life model for reinforced concrete subjected to chloride-induced corrosion, additional modelling effort will be required. The model presented in this work may be adapted for chloride ion diffusion into concrete for corrosion initiation and propagation, and for the precipitation of corrosion products within the concrete pores by the addition of mass supply and exchange terms. To simulate cracking of concrete, it is recommended that fracture mechanics be used to supplement the TPM model as was done by Ehlers and Luo [66].

A. Appendix: Linearizations

The linerization of the triphasic material model excluding the solute concentration is presented.

The balance equations are linearized according to

$$\Delta G = \frac{\partial G}{\partial \mathbf{u}_S} \Delta \mathbf{u}_S + \frac{\partial G}{\partial p^G} \Delta p^G + \frac{\partial G}{\partial p^C} \Delta p^C$$

A.1 Balance of momentum

The balance of linear momentum for the mixture is given by

$$\text{div } \mathbf{T} + \rho \mathbf{b} = 0 \tag{A.1}$$

where

$$\mathbf{T} = \mathbf{T^S} + \mathbf{T}^L + \mathbf{T}^{LQ} + \mathbf{T^G} \tag{A.2}$$

$$\rho = (1 - n^F)\, \rho^{\mathbf{S}} + (n^F\ s^{\mathbf{L}})\, \rho^L + (n^F\ s^{\mathbf{G}})\, \rho^{\mathbf{G}} \tag{A.3}$$

and $\mathbf{b}$ is the external body force. We introduce the variational formulation of the above equation as

$$G_{\mathbf{u_S}} = \int_B (\text{div } \mathbf{T} + \dot{\rho}\mathbf{b}) \cdot \delta \mathbf{u_S} \, \mathrm{d}v \tag{A.4}$$

and proceed to map to the reference configuration

$$\int_B \text{div } \mathbf{T} \, \mathrm{d}v = \int_{\partial B} \mathbf{Tn} \, \mathrm{d}a = \int_{\partial B} \mathbf{Tn} \, J_{\mathbf{S}}\, \mathbf{F}^{-T} \mathrm{d}A_{0\mathbf{S}} = \int_{\partial B} \mathbf{PN} \, \mathrm{d}A_{0\mathbf{S}} = \int_{\partial B} \text{Div } \mathbf{P} \, \mathrm{d}V_{0\mathbf{S}} \tag{A.5}$$

Using

$$\mathrm{Div}\,\mathbf{P}\cdot\delta\mathbf{u_S} = P_{ij,j}\,\delta u_i = (P_{ij}\,\delta u_i)_j - P_{ij}\,\delta u_{i,j} \tag{A.6}$$

$$\text{and} \quad \mathbf{P}\,\mathbf{N} = \mathbf{t} \tag{A.7}$$

we have

$$\int_{B_0} P_{ij}\delta u_{i,j}\,\mathrm{d}V_{0S} = \int_{\partial B_0} P_{ij}\delta u_i\,N_j\mathrm{d}A_{0S} = \int_{\partial B_0} t_i\delta u_i\,\mathrm{d}A_{0S} \tag{A.8}$$

giving

$$G_{\mathbf{us}} = \int_{B_0} \mathbf{P}\cdot\mathrm{Grad}\,\delta\mathbf{us}\,\mathrm{d}V_{0S} + \int_{B_0} J_{\mathbf{S}}\,\rho\,\mathbf{b}\cdot\delta\mathbf{us}\,\mathrm{d}V_{0S} - \int_{\partial B_0} \mathbf{t}\cdot\delta\mathbf{us}\,\mathrm{d}A_{0S} = 0 \tag{A.9}$$

Using the variation of the deformation gradient

$$\delta\mathbf{F}_S = \delta(\mathrm{Grad}\,\mathbf{u}_S + \mathbf{I}) = \mathrm{Grad}\,\delta\mathbf{u}_S \tag{A.10}$$

and the second Piola-Kirchhoff stress

$$\mathbf{P} = \mathbf{F}_S\mathbf{S} \tag{A.11}$$

we have

$$\mathbf{P}\cdot\mathrm{Grad}\,\delta\mathbf{us} = \mathbf{F}_S\mathbf{S}\cdot\delta\mathbf{F}_S. \tag{A.12}$$

With

$$F_{ij}S_{jk}\delta F_{ik} = S_{jk}F_{ij}\delta F_{ik} = \delta\mathbf{F}_S^T\mathbf{F}_S\cdot\mathbf{S} \tag{A.13}$$

and

$$\delta\mathbf{E}_S = \frac{1}{2}\delta(\mathbf{F}_S^T\mathbf{F}) = \frac{1}{2}\delta\mathbf{F}_S^T\mathbf{F}_S + \frac{1}{2}\mathbf{F}_S^T\delta\mathbf{F}_S \tag{A.14}$$

$$\delta\mathbf{E}_S = \frac{1}{2}\left[\delta\mathbf{F}_S^T\mathbf{F}_S + (\delta\mathbf{F}_S^T\mathbf{F}_S)^T\right] \tag{A.15}$$

$$\delta\mathbf{E}_S = \mathrm{sym}\,\delta\mathbf{F}_S^T\mathbf{F}_S \tag{A.16}$$

we have

$$\mathbf{P} \cdot \text{Grad } \delta\mathbf{u_S} = \mathbf{S} \cdot \delta\mathbf{E}_S \tag{A.17}$$

so that we can express the weak form of the balance of momentum for the mixture as

$$G_{\mathbf{u_S}} = \int_{B_0} \mathbf{S} \cdot \delta\mathbf{E}_S \ \delta\mathbf{u_S} \ \mathrm{d}V_{0S} + \int_{B_0} J_{\mathbf{S}} \ \rho \ \mathbf{b} \cdot \delta\mathbf{u_S} \ \mathrm{d}V_{0S} - \int_{\partial B_0} \mathbf{t} \cdot \delta\mathbf{u_S} \ \mathrm{d}A_{0S} = 0 \tag{A.18}$$

Substituting for

$$\delta\mathbf{E}_S = \frac{1}{2}\delta\mathbf{C}_S \tag{A.19}$$

into equation A.18, the final expression of the weak form of the balance of momentum for the mixture is given by

$$\mathcal{G}_{\mathbf{u_S}} = \frac{1}{2}\int_{B_0} \mathbf{S} \cdot \delta\mathbf{C} \ \mathrm{d}V_{0S} + \int_{B_0} J_S \ \rho \ \mathbf{b} \cdot \delta\mathbf{u}_S \ \mathrm{d}V_{0S} - \int_{\partial B_0} \mathbf{t} \cdot \delta\mathbf{u}_S \ \mathrm{d}A_{0S} = 0 \tag{A.20}$$

Linearization of the first term

$$\frac{1}{2}\Delta\int_{B_0} \mathbf{S} \cdot \delta\mathbf{E} \ \mathrm{d}V_{0S} = \frac{1}{2}\int_{B_0} \Delta\mathbf{S} \cdot \delta\mathbf{C} \ \mathrm{d}V_{0S} + \frac{1}{2}\int_{B_0} \mathbf{S} \cdot \Delta\delta\mathbf{C} \ \mathrm{d}V_{0S} \tag{A.21}$$

Linearization of the second Piola-Kirchhoff stress tensor, $\mathbf{S}$ with respect to the Cauchy stress $\mathbf{C}$;

$$\mathbf{S} - \mathbf{S}^F - J_S\lambda\mathbf{C}^{-1} \tag{A.22}$$

$$\Delta\mathbf{S} = \Delta\mathbf{S}^E - \Delta J_S\lambda\mathbf{C}^{-1} - J_S\Delta\lambda\mathbf{C}^{-1} - J_S\lambda\Delta\mathbf{C}^{-1} \tag{A.23}$$

$$\text{where} \quad \lambda = p^S = s^L p^L + s^G p^G. \tag{A.24}$$

$$\mathbf{S}^E = \mu^S(\mathbf{I} - \mathbf{C}^{-1}) + \lambda^S \ \log J_S \ \mathbf{C}^{-1} \tag{A.25}$$

$$\Delta\mathbf{S}^E = \frac{\partial\mathbf{S}^E}{\partial\mathbf{C}}\Delta\mathbf{C} = \left[-\mu^S\frac{\partial\mathbf{C}^{-1}}{\partial\mathbf{C}} + \lambda^S\frac{\partial\ \log J_S}{\partial\mathbf{C}}\mathbf{C}^{-1} + \lambda^S \ \log J_S \ \frac{\partial\mathbf{C}^{-1}}{\partial\mathbf{C}}\right]\Delta\mathbf{C} \tag{A.26}$$

Evaluating

$$\frac{\partial C_{ij}^{-1}}{\partial C_{on}} = -C_{io}^{-1}C_{jn}^{-1} \tag{A.27}$$

$$\frac{\partial\ \log J_S}{\partial C_{on}} = \frac{1}{2}C_{on}^{-1} \tag{A.28}$$

$$\frac{\partial\ J_S}{\partial C_{on}} = \frac{1}{2}\ J_S C_{on}^{-1} \tag{A.29}$$

yields

$$\frac{\partial S_{ij}^{E}}{\partial C_{on}}\Delta C_{on} = \left[-C_{io}^{-1}C_{jn}^{-1}(\lambda^{S}\,\log J_{S}-\mu^{S})+\frac{1}{2}\lambda^{S}C_{ij}^{-1}C_{on}^{-1}\right]\Delta C_{on} \tag{A.30}$$

Therefore

$$\frac{\partial S_{ij}}{\partial C_{on}}\Delta C_{on} = [-C_{io}^{-1}C_{jn}^{-1}(\lambda^{S}\,\log J_{S}-\mu^{S})+\frac{1}{2}\lambda^{S}C_{ij}^{-1}C_{on}^{-1}-\frac{1}{2}\lambda C_{ij}^{-1}\,J_{S}\,C_{on}^{-1}$$
$$+J_{S}\lambda C_{io}^{-1}C_{jn}^{-1}]\Delta C_{on} \tag{A.31}$$

Substituting for

$$\Delta C_{on} = 2F_{ko}\Delta F_{kn} \tag{A.32}$$

into equation A.31 yields

$$\begin{aligned}
\frac{\partial S_{ij}}{\partial C_{on}}\Delta C_{on} = &- C_{io}^{-1}C_{jn}^{-1}(\lambda^{S}\,\log J_{S}-\mu^{S})\,2F_{ko}\Delta F_{kn}\\
&+\frac{1}{2}\lambda^{S}C_{ij}^{-1}C_{on}^{-1}\,2F_{ko}\Delta F_{kn}\\
&-\frac{1}{2}\lambda C_{ij}^{-1}\,J_{S}\,C_{on}^{-1}\,2F_{ko}\Delta F_{kn}\\
&+J_{S}\lambda C_{io}^{-1}C_{jn}^{-1}\,2F_{ko}\Delta F_{kn}
\end{aligned} \tag{A.33}$$

Linearization of the second Piola-Kirchhoff stress tensor, $\mathbf{S}$ with respect to the pore pressure, λ

$$\frac{\partial \mathbf{S}}{\partial \lambda}\Delta\lambda = \Delta(-J_{S}\lambda\mathbf{C}^{-1})\Delta\lambda \tag{A.34}$$

$$\lambda = p^{S} = s^{L}p^{L}+s^{G}p^{G} \tag{A.35}$$

$$p^{S} = s^{L}(p^{G}-p^{C})+s^{G}p^{G} \tag{A.36}$$

$$p^{S} = p^{G}(s^{L}+s^{G})-s^{L}p^{C} \tag{A.37}$$

$$p^{S} = p^{G}-s^{L}p^{C} \tag{A.38}$$

Therefore

$$\frac{\partial S_{ij}}{\partial \lambda}\Delta\lambda = \frac{\partial S_{ij}}{\partial p^{G}}\Delta p^{G}+\frac{\partial S_{ij}}{\partial p^{C}}\Delta p^{C} \tag{A.39}$$

$$\frac{\partial S_{ij}}{\partial p^{G}}\Delta p^{G} = -J_{S}\,C_{ij}^{-1}\Delta p^{G} \tag{A.40}$$

$$\frac{\partial S_{ij}}{\partial p^{C}}\Delta p^{C} = s^{L}J_{S}\,C_{ij}^{-1}\Delta p^{C} \tag{A.41}$$

Hence

$$\frac{\partial \mathbf{S}}{\partial \lambda} \Delta\lambda = s^L J_S \ C_{ij}^{-1} \Delta p^C - J_S \ C_{ij}^{-1} \Delta p^G \tag{A.42}$$

and

$$\begin{aligned}
\int_{B_0} \Delta\mathbf{S} \cdot \delta\mathbf{C} \ \mathrm{d}V_{0S} = \ &\int_{B_0} -C_{io}^{-1} C_{jn}^{-1} (\lambda^S \log J_S - \mu^S) \ 2F_{ko}\delta F_{ki} \ \mathrm{d}V_{0S}\Delta F_{kn} \\
&+ \int_{B_0} \frac{1}{2}\lambda^S C_{ij}^{-1} C_{on}^{-1} \ 2F_{ko}\delta F_{ki} \ \mathrm{d}V_{0S}\Delta F_{kn} \\
&- \int_{B_0} \frac{1}{2}\lambda C_{ij}^{-1} \ J_S \ C_{on}^{-1} \ 2F_{ko}\delta F_{ki} \ \mathrm{d}V_{0S}\Delta F_{kn} \\
&+ \int_{B_0} J_S \lambda C_{io}^{-1} C_{jn}^{-1} \ 2F_{ko}\delta F_{ki} \ \mathrm{d}V_{0S}\Delta F_{kn} \\
&+ \int_{B_0} s^L J_S \ C_{ij}^{-1} \cdot \delta C_{ij} \ \mathrm{d}V_{0S}\Delta p^C \\
&- \frac{1}{2}\int_{B_0} J_S \ C_{ij}^{-1} \cdot \delta C_{ij} \ \mathrm{d}V_{0S}\Delta p^G
\end{aligned} \tag{A.43}$$

Linearization of $\delta\mathbf{C}$

$$C_{ij} = F_{ki}F_{kj} \tag{A.44}$$

$$\delta C_{ij} = \delta F_{ki}F_{kj} + F_{ki}\delta F_{kj} \tag{A.45}$$

$$\delta C_{ij} = 2\delta F_{ki}F_{kj} \tag{A.46}$$

$$\Delta\delta C_{ij} = \delta F_{ki}\Delta F_{kj} + \Delta F_{ki}\delta F_{kj} \tag{A.47}$$

$$\Delta\delta C_{ij} = 2\delta F_{ki}\Delta F_{kj} \tag{A.48}$$

Therefore

$$\frac{1}{2}\int_{B_0} S_{ij} \cdot \Delta\delta C_{ij} \ \mathrm{d}V_{0S} = \frac{1}{2}\int_{B_0} S_{ij} 2\delta F_{ki} \ \mathrm{d}V_{0S}\Delta F_{kj} = \int_{B_0} S_{ij}\delta F_{ki} \ \mathrm{d}V_{0S}\Delta F_{kj} \tag{A.49}$$

Linearization of the second term

$$\Delta \int_{B_0} J_S \ \rho \ \mathbf{b} \cdot \delta\mathbf{u}_S \ \mathrm{d}V_{0S} = \int_{B_0} \Delta J_S \ \rho \ \mathbf{b} \cdot \delta\mathbf{u}_S \ \mathrm{d}V_{0S} + \int_{B_0} J_S \ \Delta\rho \ \mathbf{b} \cdot \delta\mathbf{u}_S \ \mathrm{d}V_{0S} \tag{A.50}$$

Linearization of the Jacobian, J_S with respect to the deformation gradient, $\mathbf{F}$

$$\frac{\partial \ J_S}{\partial \ F_{on}} \Delta F_{on} = J_S \ F_{on}^{-1}\Delta F_{on} \tag{A.51}$$

Linearization of the total density ρ with respect to the gas pressure p^G

$$\rho = n^S \rho^{SR} + n^F s^L \rho^{LR} + n^F s^G \rho^{GR} \tag{A.52}$$

$$\frac{\partial \rho}{\partial p^G} \Delta p^G = \frac{\partial \rho}{\partial \rho^{GR}} \frac{\partial \rho^{GR}}{\partial p^{GR}} \Delta p^G \tag{A.53}$$

$$\frac{\partial \rho}{\partial \rho^{GR}} = n^F s^G \tag{A.54}$$

$$\rho^{GR} = \frac{p^G}{R\theta} \tag{A.55}$$

$$\frac{\partial \rho^{GR}}{\partial p^{GR}} = \frac{1}{R\theta} \tag{A.56}$$

Therefore $\quad \dfrac{\partial \rho}{\partial p^G} \Delta p^G = \dfrac{n^F s^G}{R\theta} \Delta p^G \tag{A.57}$

The specific moisture content $\frac{\partial s^L}{\partial p^C}$ is given by

$$\frac{\partial s^L}{\partial p^C} = \left(\frac{-1}{p^C}\right) jh \left[(\alpha p^C)^j \left(1 + (\alpha p^C)^j\right)\right]^{-h-1}, \tag{A.58}$$

and linearization of the Specific Moisture content with respect to the capillary pressure p^C gives

$$\frac{\partial}{\partial p^C} \left(\frac{\partial s^L}{\partial p^C}\right) \Delta p^C = \left[\frac{1}{(p^C)^2} jh \left(\alpha p^C\right)^j \left(1 + (\alpha p^C)^j\right)^{-h-2} \left(j\left(h\left(\alpha p^C\right)^j - 1\right) + 1 + (\alpha p^C)^j\right)\right] \Delta p^C. \tag{A.59}$$

Linearization of the total density ρ with respect to the capillary pressure p^C

$$\rho = n^S \rho^{SR} + n^F s^L \rho^{LR} + n^F (1 - s^L)\rho^{GR} \tag{A.60}$$

$$\frac{\partial \rho}{\partial p^C} \Delta p^C = \frac{\partial \rho}{\partial s^L} \frac{\partial s^L}{\partial p^C} \Delta p^C \tag{A.61}$$

$$\frac{\partial \rho}{\partial s^L} = n^F \rho^{LR} - n^F \rho^{GR} \tag{A.62}$$

Therefore

$$\frac{\partial \rho}{\partial p^C} \Delta p^C = (n^F \rho^{LR} - n^F \rho^{GR}) \frac{\partial s^L}{\partial p^C} \Delta p^C. \tag{A.63}$$

Hence

$$\Delta \int_{B_0} J_S \, \rho \, \mathbf{b} \cdot \delta \mathbf{u}_S \, \mathrm{d}V_{0S} = \int_{B_0} J_S \, F_{on}^{-1} \, \rho \, \mathbf{b} \cdot \delta \mathbf{u}_S \, \mathrm{d}V_{0S}\Delta F_{on}$$

$$+ \int_{B_0} J_S \, \frac{n^F s^G}{R\theta} \, \mathbf{b} \cdot \delta \mathbf{u}_S \, \mathrm{d}V_{0S}\Delta p^G \qquad (A.64)$$

$$+ \int_{B_0} J_S \, (n^F \rho^{LR} - n^F \rho^{GR}) \frac{\partial s^L}{\partial p^C} \, \mathbf{b} \cdot \delta \mathbf{u}_S \, \mathrm{d}V_{0S}\Delta p^C$$

The linearized balance of momentum of the mixture is given by

$$\Delta \mathcal{y}_{\mathbf{u}_S} = \sum_I \sum_J \delta \mathrm{d}(\mathbf{u}_S)^I \int_{B_0} \frac{\partial \mathbf{N}^I}{\partial X_i} C_{io}^{-1} C_{jn}^{-1}(-\lambda^S \, \log J_S + \mu^S) \, 2F_{kj}F_{ko} \, \frac{\partial \mathbf{N}^J}{\partial X_j} \, \mathrm{d}V_{0S}\Delta \mathrm{d}(\mathbf{u}_S)^J$$

$$+ \sum_I \sum_J \delta \mathrm{d}(\mathbf{u}_S)^I \int_{B_0} \frac{\partial \mathbf{N}^I}{\partial X_i} \lambda^S C_{ij}^{-1} C_{on}^{-1} \, F_{kj}F_{ko} \, \frac{\partial \mathbf{N}^J}{\partial X_j} \, \mathrm{d}V_{0S}\Delta \mathrm{d}(\mathbf{u}_S)^J$$

$$- \sum_I \sum_J \delta \mathrm{d}(\mathbf{u}_S)^I \int_{B_0} \frac{\partial \mathbf{N}^I}{\partial X_i} \lambda C_{ij}^{-1} \, J_S \, C_{on}^{-1} \, F_{kj}F_{ko} \, \frac{\partial \mathbf{N}^J}{\partial X_j} \, \mathrm{d}V_{0S}\Delta \mathrm{d}(\mathbf{u}_S)^J$$

$$+ \sum_I \sum_J \delta \mathrm{d}(\mathbf{u}_S)^I \int_{B_0} \frac{\partial \mathbf{N}^I}{\partial X_i} J_S\lambda C_{io}^{-1} C_{jn}^{-1} \, 2F_{kj}F_{ko} \, \frac{\partial \mathbf{N}^J}{\partial X_j} \, \mathrm{d}V_{0S}\Delta \mathrm{d}(\mathbf{u}_S)^J$$

$$+ \sum_I \sum_J \delta \mathrm{d}(\mathbf{u}_S)^I \int_{B_0} \frac{\partial \mathbf{N}^I}{\partial X_i} s^L J_S \, C_{ij}^{-1} F_{kj} \, \mathbf{N}^J \, \mathrm{d}V_{0S}\Delta \mathrm{d}(p^C)^J \qquad (A.65)$$

$$- \sum_I \sum_J \delta \mathrm{d}(\mathbf{u}_S)^I \int_{B_0} \frac{\partial \mathbf{N}^I}{\partial X_i} J_S \, C_{ij}^{-1} F_{kj} \, \mathbf{N}^J \, \mathrm{d}V_{0S}\Delta \mathrm{d}(p^G)^J$$

$$+ \sum_I \sum_J \delta \mathrm{d}(\mathbf{u}_S)^I \int_{B_0} \frac{\partial \mathbf{N}^I}{\partial X_i} J_S \, F_{on}^{-1} \, \rho \, \mathbf{b} \, \frac{\partial \mathbf{N}^J}{\partial X_n} \, \mathrm{d}V_{0S}\Delta \mathrm{d}(\mathbf{u}_S)^J$$

$$+ \sum_I \sum_J \delta \mathrm{d}(\mathbf{u}_S)^I \int_{B_0} \frac{\partial \mathbf{N}^I}{\partial X_i} J_S \, (n^F \rho^L - n^F \rho^G)\frac{\partial s^L}{\partial \mu^C} \, \mathbf{b} \, \mathbf{N}^J \, \mathrm{d}V_{0S}\Delta \mathrm{d}(p^C)^J$$

$$+ \sum_I \sum_J \delta \mathrm{d}(\mathbf{u}_S)^I \int_{B_0} \frac{\partial \mathbf{N}^I}{\partial X_i} J_S \, \frac{n^F s^G}{R\theta} \, \mathbf{b} \, \mathbf{N}^J \, \mathrm{d}V_{0S}\Delta \mathrm{d}(p^G)^J$$

A.2   Balance of mass for the liquid phase

From the balance of mass for the liquid phase,

$$(\rho^L)'_L + \rho^L \mathrm{div} \, \mathbf{x}'_L = 0 \qquad (A.66)$$

Taking the material time derivative of the liquid density with respect to the solid skeleton

$$(\rho^L)'_S = (\rho^L)'_L - \mathrm{grad} \, \rho^L \cdot \mathbf{w}_{LS}, \qquad (A.67)$$

and substituting this into equation A.66 yields

$$(\rho^L)'_S + \text{grad } \rho^L \cdot \mathbf{w}_{LS} + \rho^L \text{div } \mathbf{x}'_L = 0 \tag{A.68}$$

Recall that $\rho^L = n^L \rho^{LR}$ and $n^L = n^F s^L$ so that equation A.68 becomes

$$(n^F s^L \rho^{LR})'_S + \text{grad } n^F s^L \rho^{LR} \cdot \mathbf{w}_{LS} + n^F s^L \rho^{LR} \text{div } \mathbf{x}'_L = 0 \tag{A.69}$$

$$(n^F s^L)(\rho^{LR})'_S + (n^F \rho^{LR})(s^L)'_S + (s^L \rho^{LR})(n^F)'_S + \text{grad } n^F s^L \rho^{LR} \cdot \mathbf{w}_{LS} \\ + n^F s^L \rho^{LR} \text{div } \mathbf{x}'_L = 0 \tag{A.70}$$

Dividing through by $s^L \rho^{LR}$ and substituting for div $\mathbf{x}'_L = $ div $(\mathbf{x}'_S - \mathbf{w}'_{LS})$;

$$\frac{n^F}{\rho^{LR}}(\rho^{LR})'_S + \frac{n^F}{s^L}(s^L)'_S + (n^F)'_S + \frac{1}{s^L \rho^{LR}} \text{ grad } n^F s^L \rho^{LR} \cdot \mathbf{w}_{LS} \\ + n^F \text{div } \mathbf{x}'_S + n^F \text{div } \mathbf{w}'_{LS} = 0 \tag{A.71}$$

Summation of equation A.71 with the balance of mass for the solid phase to eliminate the $(n^F)'_S$;

$$\frac{n^F}{\rho^{LR}}(\rho^{LR})'_S + \frac{n^F}{s^L}(s^L)'_S + \frac{1}{s^L \rho^{LR}} \text{ grad } n^F s^L \rho^{LR} \cdot \mathbf{w}_{LS} + n^F \text{div } \mathbf{x}'_S + n^F \text{div } \mathbf{w}'_{LS} \\ + \frac{1 - n^F}{\rho^{SR}}(\rho^{SR})'_S + \text{div}\mathbf{x}'_S - n^F \text{div } \mathbf{x}'_S = 0 \tag{A.72}$$

Using grad $n^F s^L \rho^{LR} \cdot \mathbf{w}_{LS} = \text{div}(n^F s^L \rho^{LR} \mathbf{w}_{LS}) - n^F s^L \rho^{LR} \text{div} \mathbf{w}_{LS}$ we get;

$$\frac{n^F}{\rho^{LR}}(\rho^{LR})'_S + \frac{n^F}{s^L}(s^L)'_S + \frac{1}{s^L \rho^{LR}} \text{ div}(n^F s^L \rho^{LR} \mathbf{w}_{LS}) - n^F \text{ div}\mathbf{w}_{LS} + n^F \text{div } \mathbf{x}'_S \\ + n^F \text{div } \mathbf{w}'_{LS} + \frac{1 - n^F}{\rho^{SR}}(\rho^{SR})'_S + \text{div}\mathbf{x}'_S - n^F \text{div } \mathbf{x}'_S = 0 \tag{A.73}$$

which yields

$$\frac{n^F}{\rho^{LR}}(\rho^{LR})'_S + \frac{n^F}{s^L}(s^L)'_S + \frac{1}{s^L \rho^{LR}} \text{ div}(n^F s^L \rho^{LR} \mathbf{w}_{LS}) + \frac{1 - n^F}{(\rho^{SR})}(\rho^{SR})'_S \\ + \text{div } \mathbf{x}'_S = 0 \tag{A.74}$$

The material time derivative of the solid density in the incompressible case is given by

$$\frac{1}{\rho^{SR}}(\rho^{SR})'_S = \frac{1}{1 - n^F} \beta_S (1 - n^F) T'_S \tag{A.75}$$

and for the liquid by

$$\frac{1}{\rho^{LR}}(\rho^{LR})'_S = \frac{1}{K_L}\,(p^L)'_S - \beta_S T'_S. \tag{A.76}$$

Substituting these into equation A.74

$$\frac{n^F}{K_L}\,(p^L)'_S - n^F\,\beta_S T'_S + \frac{n^F}{s^L}(s^L)'_S + \frac{1}{s^L \rho^{LR}}\,\mathrm{div}(n^F s^L \rho^{LR}\mathbf{w}_{LS}) + \beta_S\,(1 - n^F)T'_S$$
$$+\,\mathrm{div}\,\mathbf{x}'_S = 0 \tag{A.77}$$

For an isothermal process, $T'_S = 0$;

$$\frac{n^F}{K_L}\,(p^L)'_S + \frac{n^F}{s^L}(s^L)'_S + \frac{1}{s^L \rho^{LR}}\,\mathrm{div}(n^F s^L \rho^{LR}\mathbf{w}_{LS}) + \mathrm{div}\,\mathbf{x}'_S = 0 \tag{A.78}$$

Multiplying through by s^L yields

$$\frac{n^F s^L}{K_L}\,(p^L)'_S + n^F(s^L)'_S + \frac{1}{\rho^{LR}}\,\mathrm{div}(n^F s^L \rho^{LR}\mathbf{w}_{LS}) + \frac{1}{s^L}\,\mathrm{div}\,\mathbf{x}'_S = 0 \tag{A.79}$$

Introducing the variation of capillary pressure δp^C and integrating over the differential volume gives the weak formulation of the balance of mass for the liquid phase as

$$\mathcal{G}_{p^C} = \int_{B_0} \left[\frac{n^F s^L}{K_L}\,(p^L)'_S + n^F(s^L)'_S + \frac{1}{\rho^{LR}}\,\mathrm{div}(n^F s^L \rho^{LR}\mathbf{w}_{LS}) + \frac{1}{s^L}\,\mathrm{div}\,\mathbf{x}'_S\right]\delta p^C\,\mathrm{d}v = 0 \tag{A.80}$$

Mapping to the reference configuration using the transformation equations

$$\mathrm{d}a = J_S\mathbf{F}_S^{-T}\,\mathrm{d}A_{0S} \tag{A.81}$$
$$\mathrm{d}v = J_S\,\mathrm{d}V_{0S} \tag{A.82}$$

yields

$$\mathcal{G}_{p^C} = \int_{B_0} (\frac{n^F s^L}{K_L}\,(p^L)'_S + n^F(s^L)'_S + \frac{1}{\rho^{LR}}\,\mathrm{div}(n^F s^L \rho^{LR}\mathbf{w}_{LS})$$
$$+\,\frac{1}{s^L}\,\mathrm{div}\,\mathbf{x}'_S)\delta p^C\,J_S\,\mathrm{d}V_{0S} = 0 \tag{A.83}$$

Using

$$\int_B \operatorname{div}(n^F s^L \rho^{LR} \cdot \mathbf{w}_{LS})\, \delta p^C dv = \int_{\partial B} n^F s^L \rho^{LR} \mathbf{w}_{LS} \cdot \mathbf{n}\, \delta p^C da$$
$$= \int_{\partial B_0} n^F s^L \rho^{LR} \mathbf{w}_{LS_0} \cdot \mathbf{N}\, \delta p^C dA_0 = \int_{B_0} \operatorname{Div}(n^F s^L \rho^{LR} \cdot \mathbf{w}_{LS_0})\, \delta p^C dV_{0S} \qquad (A.84)$$

where $n^F s^L\, \mathbf{w}_{LS_0} = J_S\, n^F s^L\, \mathbf{w}_{LS}\mathbf{F}_S^{-T}$ and $\operatorname{Div} \mathbf{x}'_S = \operatorname{tr}\mathbf{D}_S$. Using Gauss's theorem;

$$\operatorname{Div}\,(n^F s^L \rho^{LR}) \cdot \delta p^C = \operatorname{Div}\,(n^F s^L \rho^{LR}\delta p^C) - n^F s^L \rho^{LR}\mathbf{w}_{LS_0}\, \operatorname{Grad} \delta p^C$$

and

$$\int_{B_0} \operatorname{Div}\,(n^F s^L \rho^{LR}\delta p^C)\, dV_{0S} = \int_{\partial B_0} n^F s^L \rho^{LR}\mathbf{w}_{LS_0} \cdot \mathbf{N}\, \delta p^C\, dA_{0S}$$

we can re-write equation A.83 as

$$\mathcal{G}_{p^C} = \int_{B_0} n^F (s^L)'_S J_S\, \delta p^C\, dV_{0S} + \int_{B_0} \frac{n^F s^L}{K_L}(p^{LR})'_S J_S\, \delta p^C\, dV_{0S}$$
$$+ \int_{\partial B_0} n^F s^L \mathbf{w}_{LS_0} \cdot \mathbf{N}\, \delta p^C\, dA_{0S} - \int_{B_0} n^F s^L \mathbf{w}_{LS_0}\, \operatorname{Grad} \delta p^C\, dV_{0S} \qquad (A.85)$$
$$+ \int_{B_0} s^L J_S\, \operatorname{tr}\mathbf{D}_S\, \delta p^C\, dV_{0S} = 0$$

Substituting for the filter velocity $n^F s^L \mathbf{w}_{LS} = \frac{k k^{rL}}{\mu^L}(-\operatorname{Grad} p^{LR} + \rho^L \mathbf{g})$ into A.85;

$$\mathcal{G}_{p^C} = \int_{B_0} n^F (s^L)'_S J_S\, \delta p^C\, dV_{0S} + \int_{B_0} \frac{n^F s^L}{K_L}(p^{LR})'_S J_S\, \delta p^C\, dV_{0S}$$
$$\int_{B_0} n^F s^L \mathbf{w}_{LS_0} \cdot \mathbf{N}\, \delta p^C\, dA_{0S} - \int_{B_0} [\frac{k k^{rL}}{\mu^L}(-\operatorname{Grad} p^{LR} + \rho^L \mathbf{g})]J_S\mathbf{F}_S^{-T}\, \operatorname{Grad} \delta p^C\, dV_{0S} \quad (A.86)$$
$$+ \int_{B_0} s^L J_S\, \operatorname{tr}\mathbf{D}_S\, \delta p^C\, dV_{0S} = 0$$

Hence

$$\mathcal{G}_{p^C} = \int_{B_0} n^F (s^L)'_S J_S\, \delta p^C\, dV_{0S} + \int_{B_0} \frac{n^F s^L}{K_L}(p^{LR})'_S J_S\, \delta p^C\, dV_{0S} + \int_{B_0} n^F s^L \mathbf{w}_{LS_0} \cdot \mathbf{N}\, \delta p^C\, dA_{0S}$$
$$+ \int_{B_0} \frac{k k^{rL}}{\mu^L}\, \operatorname{Grad} p^{LR} J_S\mathbf{F}_S^{-T}\, \operatorname{Grad} \delta p^C\, dV_{0S} - \int_{B_0} \frac{k k^{rL}}{\mu^L}\, \rho^L \mathbf{g})J_S\mathbf{F}_S^{-T}\, \operatorname{Grad} \delta p^C\, dV_{0S}$$
$$+ \int_{B_0} s^L J_S\, \operatorname{tr}\mathbf{D}_S\, \delta p^C\, dV_{0S} = 0$$

$$(A.87)$$

Introducing the constitutive law

$$(s^L)'_S = \frac{\partial s^L}{\partial p^C} \frac{\partial p^C}{\partial t},$$ (A.88)

equation A.87 becomes

$$
\begin{aligned}
\mathcal{G}_{p^C} = & \int_{B_0} n^F \frac{\partial s^L}{\partial p^C} \frac{\partial p^C}{\partial t} J_S \, \delta p^C \, \mathrm{d}V_{0S} + \int_{B_0} \frac{n^F s^L}{K_L} (p^{LR})'_S J_S \, \delta p^C \, \mathrm{d}V_{0S} + \int_{B_0} n^F s^L \mathbf{w}_{LS_0} \cdot \mathbf{N} \, \delta p^C \, \mathrm{d}A_{0S} \\
& + \int_{B_0} \frac{\mathbf{k} k^{rL}}{\mu^L} \operatorname{Grad} p^{LR} J_S \mathbf{F}_S^{-T} \operatorname{Grad} \delta p^C \, \mathrm{d}V_{0S} - \int_{B_0} \frac{\mathbf{k} k^{rL}}{\mu^L} \rho^L \mathbf{g} J_S \mathbf{F}_S^{-T} \operatorname{Grad} \delta p^C \, \mathrm{d}V_{0S} \\
& + \int_{B_0} s^L J_S \, tr \mathbf{D}_S \, \delta p^C \, \mathrm{d}V_{0S} = 0
\end{aligned}
$$ (A.89)

where $\frac{\partial s^L}{\partial p^C}$ is the Specific Moisture content as previously defined. The relationship $p^L = p^G - p^C$ is employed in the linearizations.

The linearized balance of mass for the liquid phase is given by

$$
\begin{aligned}
\Delta \mathcal{G}_{p^C} =& \sum_I \sum_J \delta\mathrm{d}(p^C)^I \int_{B_0} \mathbf{N}^I n^F \frac{\partial}{\partial p^C}\left(\frac{\partial s^L}{\partial p^C}\right)(p^C)' J_S \, \mathbf{N}^J \, \mathrm{d}V_{0S} \, \Delta\mathrm{d}(p^C)^J \\
&+ \sum_I \sum_J \delta\mathrm{d}(p^C)^I \int_{B_0} \mathbf{N}^I n^F \frac{\partial s^L}{\partial p^C} \frac{\alpha}{\beta\Delta t} J_S \, \mathbf{N}^J \, \mathrm{d}V_{0S} \, \Delta\mathrm{d}(p^C)^J \\
&+ \sum_I \sum_J \delta\mathrm{d}(p^C)^I \int_{B_0} \mathbf{N}^I n^F \frac{\partial s^L}{\partial p^C}(p^C)' J_S \, \mathbf{F}_{no}^{-1} \frac{\partial \mathbf{N}^J}{\partial X_n} \, \mathrm{d}V_{0S} \, \Delta\mathrm{d}(\mathbf{u}_S)^J \\
&+ \sum_I \sum_J \delta\mathrm{d}(p^C)^I \int_{B_0} \mathbf{N}^I \frac{n^F}{K_L} \frac{\partial s^L}{\partial p^C}[(p^G)'_S - (p^C)'_S] \, J_S \, \mathbf{N}^J \, \mathrm{d}V_{0S} \, \Delta\mathrm{d}(p^C)^J \\
&+ \sum_I \sum_J \delta\mathrm{d}(p^C)^I \int_{B_0} \mathbf{N}^I \frac{n^F s^L}{K_L} \frac{\alpha}{\beta\Delta t} J_S \, \mathbf{N}^J \, \mathrm{d}V_{0S} \, \Delta\mathrm{d}(p^G)^J \\
&- \sum_I \sum_J \delta\mathrm{d}(p^C)^I \int_{B_0} \mathbf{N}^I \frac{n^F s^L}{K_L} \frac{\alpha}{\beta\Delta t} J_S \, \mathbf{N}^J \, \mathrm{d}V_{0S} \, \Delta\mathrm{d}(p^C)^J \\
&+ \sum_I \sum_J \delta\mathrm{d}(p^C)^I \int_{B_0} \mathbf{N}^I \frac{n^F s^L}{K_L}[(p^G)'_S - (p^C)'_S] \, J_S \, \mathbf{F}_{no}^{-1} \, \mathbf{N}^J \, \mathrm{d}V_{0S} \, \Delta\mathrm{d}(\mathbf{u}_S)^J \\
&+ \sum_I \sum_J \delta\mathrm{d}(p^C)^I \int_{B_0} \mathbf{N}^I \frac{\partial s^L}{\partial p^C} J_S \, (\mathbf{E}_S)'_{ij}(\mathbf{C}_S)_{ij}^{-1} \, \mathbf{N}^J \, \mathrm{d}V_{0S} \, \Delta\mathrm{d}(p^C)^J \\
&+ \sum_I \sum_J \delta\mathrm{d}(p^C)^I \int_{B_0} \mathbf{N}^I s^L J_S \, \mathbf{F}_{no}^{-1} \, (\mathbf{E}_S)'_{ij}(\mathbf{C}_S)_{ij}^{-1} \frac{\partial \mathbf{N}^J}{\partial X_n} \, \mathrm{d}V_{0S} \, \Delta\mathrm{d}(\mathbf{u}_S)^J \\
&+ \sum_I \sum_J \delta\mathrm{d}(p^C)^I \int_{B_0} \mathbf{N}^I s^L J_S \, \mathbf{F}_{no}^{-1} \frac{\alpha}{\beta\Delta t} \frac{\partial \mathbf{N}^J}{\partial X_n} \, \mathrm{d}V_{0S} \, \Delta\mathrm{d}(\mathbf{u}_S)^J \\
&- \sum_I \sum_J \delta\mathrm{d}(p^C)^I \int_{B_0} \frac{1}{2} \mathbf{N}^I s^L J_S \, (\mathbf{F}_{io}^{-1}\mathbf{F}_{jn}^{-1} + \mathbf{F}_{in}^{-1}\mathbf{F}_{jo}^{-1})\dot{\mathbf{F}}_{ik} \frac{\partial \mathbf{N}^J}{\partial X_n} \, \mathrm{d}V_{0S} \, \Delta\mathrm{d}(\mathbf{u}_S)^J \\
&+ \sum_I \sum_J \delta\mathrm{d}(p^C)^I \int_{B_0} \frac{\partial \mathbf{N}^I}{\partial X_j} \frac{k k^{rL}}{\mu^L} J_S \, \mathbf{F}_{ij}^{-T} \frac{\partial \mathbf{N}^J}{\partial X_n} \, \mathrm{d}V_{0S} \, \Delta\mathrm{d}(p^G)^J \\
&+ \sum_I \sum_J \delta\mathrm{d}(p^C)^I \int_{B_0} \frac{\partial \mathbf{N}^I}{\partial X_j} \frac{k k^{rL}}{\mu^L} J_S \, \mathbf{F}_{no}^{-1}\mathbf{F}_{ij}^{-T} \, \mathrm{Grad}p^L \frac{\partial \mathbf{N}^J}{\partial X_n} \, \mathrm{d}V_{0S} \, \Delta\mathrm{d}(\mathbf{u}_S)^J \\
&- \sum_I \sum_J \delta\mathrm{d}(p^C)^I \int_{B_0} \frac{\partial \mathbf{N}^I}{\partial X_j} \frac{k k^{rL}}{\mu^L} J_S \, \mathbf{F}_{jo}^{-1}\mathbf{F}_{ni}^{-1} \, \mathrm{Grad}p^L \frac{\partial \mathbf{N}^J}{\partial X_n} \, \mathrm{d}V_{0S} \, \Delta\mathrm{d}(\mathbf{u}_S)^J \\
&- \sum_I \sum_J \delta\mathrm{d}(p^C)^I \int_{B_0} \frac{\partial \mathbf{N}^I}{\partial X_j} \frac{k k^{rL}}{\mu^L} J_S \, \mathbf{F}_{ij}^{-T} \frac{\partial \mathbf{N}^J}{\partial X_n} \, \mathrm{d}V_{0S} \, \Delta\mathrm{d}(p^C)^J \\
&+ \sum_I \sum_J \delta\mathrm{d}(p^C)^I \int_{B_0} \frac{\partial \mathbf{N}^I}{\partial X_j} \frac{k}{\mu^L} \frac{\partial k^{rL}}{\partial s^L} \frac{\partial s^L}{\partial p^C} J_S \, \mathbf{F}_{ij}^{-T} \, \mathrm{Grad}p^L \mathbf{N}^J \, \mathrm{d}V_{0S} \, \Delta\mathrm{d}(p^C)^J \\
&- \sum_I \sum_J \delta\mathrm{d}(p^C)^I \int_{B_0} \mathbf{N}^I \frac{\partial s^L}{\partial p^C} \frac{\partial p^C}{\partial t} \, \mathbf{F}_{no}^{-1} \frac{\partial \mathbf{N}^J}{\partial X_n} \, \mathrm{d}V_{0S} \, \Delta\mathrm{d}(\mathbf{u}_S)^J \\
&- \sum_I \sum_J \delta\mathrm{d}(p^C)^I \int_{B_0} \mathbf{N}^I \frac{s^L}{K_L}(p^L)'_S \, \mathbf{F}_{no}^{-1} \frac{\partial \mathbf{N}^J}{\partial X_n} \, \mathrm{d}V_{0S} \, \Delta\mathrm{d}(\mathbf{u}_S)^J
\end{aligned}
\tag{A.90}
$$

A.3 Balance of mass for the gas phase

From the balance of mass of the gas phase

$$(\rho^G)'_G + \rho^G \operatorname{div} \mathbf{x}'_G = 0 \tag{A.91}$$

Taking it's material time derivative with respect to the solid skeleton

$$(\rho^G)'_G = (\rho^G)'_S + \operatorname{grad} \rho^G \cdot \mathbf{w}_{GS} \tag{A.92}$$

and substituting this into equation A.91;

$$(\rho^G)'_S + \operatorname{grad} \rho^G \cdot \mathbf{w}_{GS} + \rho^G \operatorname{div} \mathbf{x}'_G = 0 \tag{A.93}$$

Recall that $\rho^G = n^G \rho^{GR}$ and $n^G = n^F s^G$ so that equation A.93 becomes

$$(n^F s^G \rho^{GR})'_S + \operatorname{grad} (n^F s^G \rho^{GR}) \cdot \mathbf{w}_{GS} + (n^F s^G \rho^{GR}) \operatorname{div} \mathbf{x}'_G = 0 \tag{A.94}$$

Expanding equation A.94;

$$(n^F s^G)(\rho^{GR})'_S + (n^F \rho^{GR})(s^G)'_S + (s^G \rho^{GR})(n^F)'_S + \operatorname{grad} (n^F s^G \rho^{GR}) \cdot \mathbf{w}_{GS} + (n^F s^G \rho^{GR}) \operatorname{div} \mathbf{x}'_G = 0 \tag{A.95}$$

Dividing through by $s^G \rho^{GR}$ and substituting for $\operatorname{div} \mathbf{x}'_G = \operatorname{div} (\mathbf{x}'_S + \mathbf{w}_{GS})$;

$$\begin{aligned}
&\frac{n^F}{\rho^{GR}}(\rho^{GR})'_S + \frac{n^F}{s^G}(s^G)'_S + (n^F)'_S + \frac{1}{s^G \rho^{GR}} \operatorname{grad} (n^F s^G \rho^{GR}) \cdot \mathbf{w}_{GS} \\
&+ n^F \operatorname{div} (\mathbf{x}'_S + \mathbf{w}_{GS}) = 0
\end{aligned} \tag{A.96}$$

Summation of equation A.96 with the balance of mass for the solid skeleton to eliminate $(n^F)'_S$;

$$\begin{aligned}
&\frac{n^F}{\rho^{GR}}(\rho^{GR})'_S + \frac{n^F}{s^G}(s^G)'_S + \frac{1}{s^G \rho^{GR}} \operatorname{grad} (n^F s^G \rho^{GR}) \cdot \mathbf{w}_{GS} + n^F \operatorname{div} \mathbf{x}'_S \\
&+ n^F \operatorname{div} \mathbf{w}_{GS} + \frac{1 - n^F}{\rho^{SR}}(\rho^{SR})'_S + \operatorname{div} \mathbf{x}'_S - n^F \operatorname{div} \mathbf{x}'_S = 0
\end{aligned} \tag{A.97}$$

Using grad $(n^F s^F \rho^{GR}) \cdot \mathbf{w}_{GS} = \mathrm{div}\ (n^F s^F \rho^{GR} \mathbf{w}_{GS}) - (n^F s^F \rho^{GR})\mathrm{div}\ \mathbf{w}_{GS}$ we get;

$$\frac{n^F}{\rho^{GR}}(\rho^{GR})'_S + \frac{n^F}{s^G}(s^G)'_S + \frac{1}{s^G \rho^{GR}}\mathrm{div}\ (n^F s^F \rho^{GR}\mathbf{w}_{GS}) - n^F \mathrm{div}\ \mathbf{w}_{GS}$$
$$+ n^F\ \mathrm{div}\ \mathbf{x}'_S + n^F\ \mathrm{div}\ \mathbf{w}_{GS} + \frac{1-n^F}{\rho^{SR}}(\rho^{SR})'_S + \mathrm{div}\ \mathbf{x}'_S - n^F\ \mathrm{div}\ \mathbf{x}'_S = 0 \tag{A.98}$$

which yields

$$\frac{n^F}{\rho^{GR}}(\rho^{GR})'_S + \frac{n^F}{s^G}(s^G)'_S + \frac{1}{s^G \rho^{GR}}\mathrm{div}\ (n^F s^F \rho^{GR}\mathbf{w}_{GS}) + \frac{1-n^F}{\rho^{SR}}(\rho^{SR})'_S + \mathrm{div}\ \mathbf{x}'_S = 0 \quad \text{(A.99)}$$

The material time derivative of the solid density in the incompressible case is given in equation A.75 and for the gas phase is given by

$$(\rho^{GR})'_S = \frac{1}{R\theta}(p^G)'_S \tag{A.100}$$

Substituting these into equation A.99;

$$\frac{n^F}{\rho^{GR} R\theta}(p^G)'_S + \frac{n^F}{s^G}(s^G)'_S + \frac{1}{s^G \rho^{GR}}\mathrm{div}\ (n^F s^F \rho^{GR}\mathbf{w}_{GS}) + \beta_S\ (1 - n^F)T'_S + \mathrm{div}\ \mathbf{x}'_S = 0 \quad \text{(A.101)}$$

For an isothermal process, $T'_S = 0$ so that equation A.101 becomes

$$\frac{n^F}{\rho^{GR} R\theta}(p^G)'_S + \frac{n^F}{s^G}(s^G)'_S + \frac{1}{s^G \rho^{GR}}\mathrm{div}\ (n^F s^F \rho^{GR}\mathbf{w}_{GS}) + \mathrm{div}\ \mathbf{x}'_S = 0 \tag{A.102}$$

Multiplying through by s^G;

$$\frac{n^F s^G}{\rho^{GR} R\theta}(p^G)'_S + n^F(s^G)'_S + \frac{1}{\rho^{GR}}\mathrm{div}\ (n^F s^F \rho^{GR}\mathbf{w}_{GS}) + s^G \mathrm{div}\ \mathbf{x}'_S = 0 \tag{A.103}$$

Hence the weak form of the balance of mass of the gas phase is given by

$$\mathcal{G}_{p^G} = \int_B \left[\frac{n^F s^G}{\rho^{GR} R\theta}(p^G)'_S + n^F(s^G)'_S + \frac{1}{\rho^{GR}}\mathrm{div}\ (n^F s^F \rho^{GR}\mathbf{w}_{GS}) + s^G \mathrm{div}\ \mathbf{x}'_S \right] \delta p^G\ \mathrm{d}v = 0 \quad \text{(A.104)}$$

Mapping to the reference configuration equation A.81

$$\mathcal{G}_{p^G} = \int_{B_0} \left[\frac{n^F s^G}{\rho^{GR} R\theta}(p^G)'_S + n^F(s^G)'_S + \frac{1}{\rho^{GR}}\mathrm{div}\ (n^F s^F \rho^{GR}\mathbf{w}_{GS}) \right.$$
$$\left. + s^G \mathrm{div}\ \mathbf{x}'_S \right] \delta p^G\ J_S \mathrm{d}V = 0 \tag{A.105}$$

and making use of the relationship in A.84 and Gauss' theorem we obtain;

$$\mathcal{G}_{p^G} = \int_{B_0} \frac{n^F s^G}{\rho^G R\theta} (p^G)'_S \, \delta p^G \, dV_{0S} + \int_{B_0} n^F (s^G)'_S J_S \, \delta p^G \, dV_{0S} + \int_{B_0} n^F s^G \mathbf{w}_{GS} \cdot \mathbf{N} \, \delta p^G \, dA_{0S}$$
$$- \int_{B_0} n^F s^G \mathbf{w}_{GS} \, \mathrm{Grad} \, \delta p^G \, dV_{0S} + \int_{B_0} s^G J_S \, \mathrm{tr} \mathbf{D}_S \, \delta p^G \, dV_{0S} = 0$$

$$(A.106)$$

Substituting for the filter velocity $n^F s^G \mathbf{w}_{GS} = \frac{\mathbf{k} k^{rG}}{\mu^G}(-\mathrm{Grad}\, p^G + \rho^G \mathbf{g})$ into the above equation yields

$$\mathcal{G}_{p^G} = \int_{B_0} \frac{n^F s^G}{\rho^G R\theta} (p^G)'_S \, \delta p^G \, dV_{0S} + \int_{B_0} n^F (s^G)'_S J_S \, \delta p^G \, dV_{0S} + \int_{B_0} n^F s^G \mathbf{w}_{GS} \cdot \mathbf{N} \, \delta p^G \, dA_{0S}$$
$$- \int_{B_0} \frac{\mathbf{k} k^{rG}}{\mu^G}(-\mathrm{Grad}\, p^G + \rho^G \mathbf{g}) \, \mathrm{Grad} \, \delta p^G \, dV_{0S} + \int_{B_0} s^G J_S \, \mathrm{tr} \mathbf{D}_S \, \delta p^G \, dV_{0S} = 0$$

$$(A.107)$$

which may be re-arranged to give

$$\mathcal{G}_{p^G} = \int_{B_0} \frac{n^F s^G}{\rho^G R\theta} (p^G)'_S \, \delta p^G \, dV_{0S} + \int_{B_0} n^F (s^G)'_S J_S \, \delta p^G \, dV_{0S} + \int_{B_0} n^F s^G \mathbf{w}_{GS} \cdot \mathbf{N} \, \delta p^G \, dA_{0S}$$
$$+ \int_{B_0} \frac{\mathbf{k} k^{rG}}{\mu^G} \mathrm{Grad} \, p^G \, \mathrm{Grad} \, \delta p^G \, dV_{0S} - \int_{B_0} \frac{\mathbf{k} k^{rG}}{\mu^G} \rho^G \mathbf{g} \, \mathrm{Grad} \, \delta p^G \, dV_{0S}$$
$$+ \int_{B_0} s^G J_S \, \mathrm{tr} \mathbf{D}_S \, \delta p^G \, dV_{0S} = 0$$

$$(A.108)$$

Using $s^G = 1 - s^L$ and $(s^G)'_S = -(s^L)'_S$ and making use of the constitutive law given in A.88, we obtain

$$\mathcal{G}_{p^G} = \int_{B_0} \frac{n^F s^G}{\rho^G R\theta} (p^G)'_S \, \delta p^G \, dV_{0S} - \int_{B_0} n^F \frac{\partial s^L}{\partial p^C} \frac{\partial p^C}{\partial t} J_S \, \delta p^G \, dV_{0S} + \int_{B_0} n^F s^G \mathbf{w}_{GS} \cdot \mathbf{N} \, \delta p^G \, dA_{0S}$$
$$+ \int_{B_0} \frac{\mathbf{k} k^{rG}}{\mu^G} \mathrm{Grad} \, p^G \, \mathrm{Grad} \, \delta p^G \, dV_{0S} - \int_{B_0} \frac{\mathbf{k} k^{rG}}{\mu^G} \rho^G \mathbf{g} \, \mathrm{Grad} \, \delta p^G \, dV_{0S}$$
$$+ \int_{B_0} s^G J_S \, \mathrm{tr} \mathbf{D}_S \, \delta p^G \, dV_{0S} = 0$$

$$(A.109)$$

The linearized balance of mass of the gas phase is given by

$$
\begin{aligned}
\Delta \mathcal{G}_{p^G} = &- \sum_I \sum_J \delta\mathrm{d}(p^G)^I \int_{B_0} \mathbf{N}^I n^F \frac{\partial}{\partial p^C}\Big(\frac{\partial s^L}{\partial p^c}\Big)(p^C)' \, J_S \, \mathbf{N}^J \mathrm{d}V_{0S} \, \Delta\mathrm{d}(p^C)^J \\
&- \sum_I \sum_J \delta\mathrm{d}(p^G)^I \int_{B_0} \mathbf{N}^I n^F \frac{\partial s^L}{\partial p^c}\frac{\alpha}{\beta\Delta t} \, J_S \, \mathbf{N}^J \mathrm{d}V_{0S} \, \Delta\mathrm{d}(p^C)^J \\
&- \sum_I \sum_J \delta\mathrm{d}(p^G)^I \int_{B_0} \mathbf{N}^I n^F \frac{\partial s^L}{\partial p^c}(p^C)' \, J_S \mathbf{F}_{no}^{-1} \frac{\partial \mathbf{N}^J}{\partial X_n}\mathrm{d}V_{0S} \, \Delta\mathrm{d}(\mathbf{u}_S)^J \\
&- \sum_I \sum_J \delta\mathrm{d}(p^G)^I \int_{B_0} \mathbf{N}^I \frac{n^F}{\rho^G R\theta}(p^G)'_S \frac{\partial s^L}{\partial p^C} \, J_S \, \mathbf{N}^J \mathrm{d}V_{0S} \, \Delta\mathrm{d}(p^C)^J \\
&+ \sum_I \sum_J \delta\mathrm{d}(p^G)^I \int_{B_0} \mathbf{N}^I \frac{n^F}{\rho^G R\theta}\frac{\alpha}{\beta\Delta t} \, J_S \, \mathbf{N}^J \mathrm{d}V_{0S} \, \Delta\mathrm{d}(p^G)^J \\
&- \sum_I \sum_J \delta\mathrm{d}(p^G)^I \int_{B_0} \mathbf{N}^I \frac{n^F s^F}{R\theta}(p^G)'_S \frac{(\rho^G)^{-2}}{R\theta} \, J_S \, \mathbf{N}^J \mathrm{d}V_{0S} \, \Delta\mathrm{d}(p^G)^J \\
&+ \sum_I \sum_J \delta\mathrm{d}(p^G)^I \int_{B_0} \mathbf{N}^I \frac{n^F s^G}{\rho^G R\theta}(p^G)'_S \, J_S \, \mathbf{F}_{no}^{-1} \frac{\partial \mathbf{N}^J}{\partial X_n}\mathrm{d}V_{0S} \, \Delta\mathrm{d}(\mathbf{u}^S)^J \\
&- \sum_I \sum_J \delta\mathrm{d}(p^G)^I \int_{B_0} \mathbf{N}^I \frac{\partial s^L}{\partial p^C} \, J_S \, (\mathbf{E}_S)'_{ij}(\mathbf{C}_S)_{ij}^{-1} \, \mathbf{N}^J \mathrm{d}V_{0S} \, \Delta\mathrm{d}(p^C)^J \\
&+ \sum_I \sum_J \delta\mathrm{d}(p^G)^I \int_{B_0} \mathbf{N}^I s^G \, J_S\mathbf{F}_{no}^{-1} \, (\mathbf{E}_S)'_{ij}(\mathbf{C}_S)_{ij}^{-1} \frac{\partial \mathbf{N}^J}{\partial X_n}\mathrm{d}V_{0S} \, \Delta\mathrm{d}(\mathbf{u}_S)^J \qquad (A.110)\\
&+ \sum_I \sum_J \delta\mathrm{d}(p^G)^I \int_{B_0} \mathbf{N}^I s^G \, J_S\mathbf{F}_{no}^{-1} \frac{\alpha}{\beta\Delta t} \frac{\partial \mathbf{N}^J}{\partial X_n}\mathrm{d}V_{0S} \, \Delta\mathrm{d}(\mathbf{u}_S)^J \\
&- \sum_I \sum_J \delta\mathrm{d}(p^G)^I \int_{B_0} \frac{1}{2}\mathbf{N}^I s^G \, J_S(\mathbf{F}_{io}^{-1}\mathbf{F}_{jn}^{-1} + \mathbf{F}_{in}^{-1}\mathbf{F}_{jo}^{-1})\dot{\mathbf{F}}_{ij} \frac{\partial \mathbf{N}^J}{\partial X_n}\mathrm{d}V_{0S} \, \Delta\mathrm{d}(\mathbf{u}_S)^J \\
&+ \sum_I \sum_J \delta\mathrm{d}(p^G)^I \int_{B_0} \frac{\partial \mathbf{N}^I}{\partial X_j}\frac{\mathbf{k}k^{rL}}{\mu^G} \, J_S \, \mathbf{F}_{ij}^{-T} \frac{\partial \mathbf{N}^J}{\partial X_n}\mathrm{d}V_{0S} \, \Delta\mathrm{d}(p^G)^J \\
&+ \sum_I \sum_J \delta\mathrm{d}(p^G)^I \int_{B_0} \frac{\partial \mathbf{N}^I}{\partial X_j}\frac{\mathbf{k}k^{rL}}{\mu^G} \, J_S \, \mathbf{F}_{no}^{-1}\mathbf{F}_{ij}^{-T} \, \mathrm{Grad}p^G \frac{\partial \mathbf{N}^J}{\partial X_n}\mathrm{d}V_{0S} \, \Delta\mathrm{d}(\mathbf{u}_S)^J \\
&- \sum_I \sum_J \delta\mathrm{d}(p^G)^I \int_{B_0} \frac{\partial \mathbf{N}^I}{\partial X_j}\frac{\mathbf{k}k^{rL}}{\mu^G} \, J_S \, \mathbf{F}_{jo}^{-1}\mathbf{F}_{ni}^{-1} \, \mathrm{Grad}p^G \frac{\partial \mathbf{N}^J}{\partial X_n}\mathrm{d}V_{0S} \, \Delta\mathrm{d}(\mathbf{u}_S)^J \\
&+ \sum_I \sum_J \delta\mathrm{d}(p^G)^I \int_{B_0} \frac{\partial \mathbf{N}^I}{\partial X_j}\frac{\mathbf{k}}{\mu^G}\frac{\partial k^{rG}}{\partial s^L}\frac{\partial s^L}{\partial p^C} \, J_S \, \mathbf{F}_{ij}^{-T} \, \mathrm{Grad}p^G \frac{\partial \mathbf{N}^J}{\partial X_n} \, \mathrm{d}V_{0S} \, \Delta\mathrm{d}(p^C)^J \\
&+ \sum_I \sum_J \delta\mathrm{d}(p^G)^I \int_{B_0} \mathbf{N}^I n^F \frac{\partial s^L}{\partial p^c}(p^C)' \, \mathbf{F}_{no}^{-1} \frac{\partial \mathbf{N}^J}{\partial X_n}\mathrm{d}V_{0S} \, \Delta\mathrm{d}(\mathbf{u}_S)^J \\
&- \sum_I \sum_J \delta\mathrm{d}(p^G)^I \int_{B_0} \mathbf{N}^I \frac{s^G}{\rho^G R\theta}(p^G)'_S \, J_S\mathbf{F}_{no}^{-1} \frac{\partial \mathbf{N}^J}{\partial X_n}\mathrm{d}V_{0S} \, \Delta\mathrm{d}(\mathbf{u}_S)^J
\end{aligned}
$$

B. Appendix: Ethics approval

APPLICATION FORM

APPLICANT'S DETAILS		
Name of principal researcher, student or external applicant		Joanitta Ndawula
Department		Civil engineering
Preferred email address of applicant:		ndwjoa001@myuct.ac.za
	Your Degree: e.g., MSc, PhD, etc.,	MSc
	Name of Supervisor (if supervised):	Dr Sebastian Skatulla
If this is a researchcontract, indicate the source of funding/sponsorship		
Project Title		Multiphase modelling of the deterioration of reinforced concrete structures

-
-
-
-
-
-

SIGNED BY	Full name	Signature	Date
Principal Researcher/ Student/External applicant	Joanitta Ndawula		13 October 2016

APPLICATION APPROVED BY	Full name	Signature	Date
Supervisor (where applicable)	Dr Sebastian Skatulla		13/10/16
HOD (or delegated nominee) Final authority for all applicants who have answered NO to all questions in Section1; and for all Undergraduateresearch (Including Honours).	M Vandersche ~~Professor Neil Armitage~~		31/10/16
Chair : Faculty EIR Committee For applicants other than undergraduate students who have answered YES to any of the above questions.			

References

[1] K. Toongoenthong and K. Maekawa. Simulation of coupled corrosive product formation, migration into crak and propagation in reinforced concrete sections. *Journal of Advanced Concrete Technology*, 3(2):253–265, 2005.

[2] D.V. Val, L. Chernin, and M.G. Stewart. Experimental and numerical investigation of corrosion-induced cover cracking in reinforced concrete structures. *Journal of Structural Engineering*, 135(4):376–385, 2009.

[3] Alexander Michel, Brad J Pease, Adéla Peterová, Mette R Geiker, Henrik Stang, and Anna A Emilie Thybo. Penetration of corrosion products and corrosion-induced cracking in reinforced cementitious materials: Experimental investigations and numerical simulations. *Cement and Concrete Composites*, 47(75-86), 2014.

[4] P. K. Mehta and P. J. M. Monteiro. *Concrete: microstructure, properties, and materials.* McGraw-Hill, 3rd edition, 2006.

[5] Yunus Ballim, Mark Alexander, and Hans Beushausen. *Fulton's concrete technology*, chapter 9. Durability of concrete, pages 155–188. Cement and Concrete Institute, 9 edition, 2009.

[6] M G Alexander and H Beushausen. Performance-based durability testing, design and specification in south africa: latest developments. In Mukesh C Limbachiya and Hsein Y. Kew, editors, *Excellence in Concrete Construction through Innovation*, pages 429–434, 2008.

[7] Ueli Angst, Bernhard Elsener, Claus K. Larsen, and Øystein Vennesland. Critical chloride content in reinforced concrete — a review. *Cement and Concrete Research*, 39(12):1122–1138, 2009.

[8] Qiang Yuan, Caijun Shi, Geert De Schutter, Dehua Deng, and Fuqiang He. Numerical model for chloride penetration into saturated concrete. *Journal of Materials in Civil Engineering*, 23(3):305–311, 2011.

[9] Mike Benjamin Otieno. *The Development of Empirical Chloride-induced Corrosion Rate Pre-diction Models for Cracked and Uncracked Steel Reinforced Concrete Structures in the Marine Tidal Zone*. PhD book, University of Cape Town, February 2014.

[10] Yuxi Zhao, Yingyao Wu, and Jin Weiliang. Distribution of millscale on corroded steel bars and penetration of steel corrosion products in concrete. *Corrosion Science*, 66:160–168, 2013.

[11] Y. Zhou, B. Gencturk, K. William, and A. Attar. Carbonation-induced and chloride-induced corrosion in reinforced concrete structures. *Journal of Materials in Civil Engineering*, 27(9), 2015.

[12] A. Ghani Isgor, O. Burkan Razaqpur. Modelling steel corrosion in concrete structures. *Materials and Structures*, 39(3):291–302, 2007.

[13] Y. Liu and R. E. Weyers. Modeling the time to corrosion cracking in chloride contaminated reinforced concrete structures. *ACI Materials Journal*, 95(6):675–680, 1998.

[14] Jing Hu. *Porosity of Concrete: Morphological Study of Model Concrete*. PhD book, Delft University of Technology, October 2004.

[15] Jacob Bear and Yehuda Bachmat. *Introduction to modeling of transport phenomena in porous media*. Kluwer Academic Publishers, 1st edition, 1990.

[16] Reint de Boer. *Theory of Porous Media: Highlights in Historical Development and Current State*. Springer-Verlag Berlin Heidelberg, 1 edition, 2000. doi: 10.1007/978-3-642-59637-7.

[17] G. Hopkins, S. Skatulla, L. Moj, T. Ricken, N. Ntusi, and E. Meintjes. A biphasic model for full cycle simulation of the human heart aimed at rheumatic heart disease. *Computers & Structures*, 2018.

[18] Tim Ricken, Andrea Sindern, Joachim Bluhm, Renatus Widmann, Martin Denecke, Tobias Gehrke, and Torsten C Schmidt. Concentration driven phase transitions in multiphase porous media with application to methane oxidation in landfill cover layers. *ZAMM Journal of applied mathematics and mechanics: Zeitschrift für angewandte Mathematik und Mechanik*, pages 1–14, 2013.

[19] Gary Hopkins. Growth, modelling and remodelling of cardiac tissue: A multiphase approach requirements. mathesis, University of Cape Town, 2017.

[20] D Gawin, F Pesavento, and B A Schrefler. Modelling of hygro - thermal behaviour of concrete at high temperature with thermo - chemical and mechanical material degradation. *Computer Methods in Applied Mechanics and Engineering*, 192:1731–1771, 2003.

[21] D Gawin, M Koniorczyk, and F Pesavento. Modelling of hydro-thermo-chemo-mechanical phenomena in building materials. *Bulletin Of The Polish Academy Of Sciences Technical Sciences*, 61(1):51–63, 2013.

[22] F Pesavento, B A. Schrefler, and G Sciumè. Multiphase flow in deforming porous media: A review. *Archives of Computational Methods in Engineering*, pages 1–26, 2016. doi: 10.1007/s11831-016-9171-6.

[23] G. Sciumè, F. Benboudjema, C. De Sa, F. Pesavento, Y. Berthaud, and B.A. Schrefler. A multiphysics model for concrete at early age applied to repair problems. *Engineering Structures*, 57:374–387, 2013.

[24] V.A. Salomoni, C.E. Majorana, Mazzucco G., G. Xotta, and G.A. Khoury. *Multiscale modelling of Concrete as a fully coupled porous medium*, pages 171–231. 2009.

[25] Steinar Helland. Design for service life: Implementation of fib model code 2010 rules in the operational code iso 16204. *Structural Concrete*, 14(1):10–18, 2013.

[26] I. Balafas and C. J. Burgoyne. Environmental effects on cover cracking due to corrosion. *Cement and Concrete Research*, 40(9):1429–1440, 2010.

[27] Raja Rizwan Hussain and Tetsuya Ishida. Critical carbonation depth for initiation of steel corrosion in fully carbonated concrete and development of electrochemical carbonation induced corrosion model. *International Journal of Electrochemical Science*, 4:1178 – 1195, August 2009.

[28] F. Bouchaala, C. Payan, V. Garnier, and J. P. Balayssac. Carbonation assessment in concrete by nonlinear ultrasound. *Cement and Concrete Research*, 41:557–559, 2011.

[29] M. F. Montemor, A. M. P. Simoes, and M. G. S. Ferreira. Chloride-induced corrosion on reinforcing steel: from the fundamentals to the monitoring techniques. *Cement & Concrete Composites*, 25:491–502, May–July 2003.

[30] K. L. Scrivener. *Materials Science of Concrete*, chapter The microstructure of concrete. The American Ceramic Society, Westerville, OH, USA, 1989.

[31] D.M. Roy, G.M. Idorn, and Strategic Highway Research Program (U.S.). *Concrete Microstructure*. SHRP Reports. Strategic Highway Research Program, National Research Council, 1993. ISBN 9780309052542.

[32] Ameer A. Hilal. *High Performance Concrete Technology and Applications*, chapter Microstructure of Concrete, pages 3–24. IntechOpen, 2016. URL https://www.intechopen.com/books/high-performance-concrete-technology-and-applications/microstructure-of-concrete.

[33] Sidney Diamond. The microstructure of cement paste and concrete—a visual primer. *Cement & Concrete Composites*, 26(8):919–933, November 2004.

[34] Azadeh Attari, Ciaran McNally, and Mark G. Richardson. A combined sem-calorimetric approach for assessing hydration and porosity development in ggbs concrete. *Cement and Concrete Composites*, 68:46–56, 2016.

[35] D. N. Winslow and M. D. Cohen. Percolation and pore structure in mortars and concrete. *Cement and Concrete Research*, 24(1):25–37, 1994.

[36] Yingshu Yuan and Yongsheng Ji. Modeling corroded section configuration of steel bar in concrete structure. *Construction and Building Materials*, 23(6):2461–2466, 2009.

[37] R.W Lewis and B.A Schrefler. *The finite element method in the static and dynamic deformation and consolidation of porous media*. John WIley & Sons, Ltd. (UK), 2 edition, 1998.

[38] Wolfgang Ehlers and Joachim Bluhm, editors. *Porous Media: Theory, Experiments and Numerical Applications*. Springer, 2002.

[39] HL Toor and SH Chiang. Diffusion-controlled chemical reactions. *AIChE Journal*, 5(3):339–344, 1959.

[40] M Pérez, B Pantazopoulou, and S J ; A Thomas. Numerical solution of mass transport equations in concrete structures. *Computers & Structures*, 79(13):1251–1264, 2001.

[41] Yury A. Villagran Zaccardi, Natalia M. Alderete, and Nele De Belie. Improved model for capillary absorption in cementitious materials: Progress over the fourth root of time. *Cement and Concrete Research*, 100:153–165, 2017.

[42] Nicos S. Martys and Chiara F. Ferraris. Capillary transpport in mortars and concrete. *Cement and Concrete Research*, 27(5):747–760, 1997.

[43] S. Dehghanpoor Abyaneh, H.S. Wong, and N.R. Buenfeld. Computational investigation of capillary absorption in concrete using a three-dimensional mesoscale approach. *Computational Materials Science*, 87:54–64, 2014.

[44] John R. Fanchi. *Principles of Applied Reservoir Simulation*. Gulf Professional Publishing, 4th edition, 2018.

[45] Ralph H. Petrucci, Geoffrey Herring, Jeffrey D. Madura, and Carey Bissonnette. *General chemistry : principles and modern applications*. Pearson Canada Inc, 11th edition, 2017.

[46] M. T. van Genuchten. A closed-form equation for predicting the hydraulic conductivity of unsaturated soils. *Soil Science Society of America Journal*, 44(5):892–898, 1980.

[47] P.J. Van Geel and S.D. Roy. A proposed model to include a residual napl saturation in a hysteretic capillary pressure–saturation relationship. *Journal of Contaminant Hydrology*, 58: 79–110, 2002.

[48] J.P. Monlouis-Bonnaire, J. Verdier, and B. Perrin. Prediction of relative permeability to gas flow of cement-based materials. *Cement and concrete research*, 34:737–744, 2004.

[49] R.H. Brooks and A.T. Corey. Properties of porous media affecting fluid flow. *Journal of the Irrigation and Drainage Division*, 92(2):61–90, 1966.

[50] D Gawin, C.E. Majorana, and B A Schrefler. Numerical analysis of hygro-thermal behaviour and damage of concrete at high temperature. *Mechanics of cohesive-frictional materials*, 4: 37–74, 1999.

[51] P. Haupl, J. Grunewald, and H. Fechner. Coupled heat air and moisture transfer in building structures. *International Journal of Heat and Mass Transfer*, 40(7):1633–1642, 1997.

[52] Ueli M. Angst, Mette R. Geiker, Alexander Michel, Christoph Gehlen, Hong Wong, O. Burkan Isgor, Bernhard Elsener, Carolyn M. Hansson, Raoul François, Karla Hornbostel, Rob Polder, Maria Cruz Alonso, Mercedes Sanchez, Maria João Correia, Maria Criado, A. Sagüés, and Nick Buenfeld. The steel–concrete interface. *Materials and Structures*, 50(143), 2017.

[53] Göran Fagerlund. *A service life model for internal frost damage in concrete*. PhD book, Lund University, 2004.

[54] Biologydictionary.net Editors. Water potential, 2017. URL `https://biologydictionary.net/water-potential/`.

[55] Martin Schneebeli, Hannes Fluhler, Thomas Gimmi, Hannes Wydler, Hans-Peter Laeser, and Toni Baer. Measurements of water potential and water content in unsaturated crystalline rock. *WATER RESOURCES RESEARCH*, 31(8):1837–1843, 1995.

[56] R. W. Revie and H. H. Uhlig. *Corrosion and corrosion control. An Introduction to Corrosion Science and Engineering.* John WIley & Sons, Inc, 2008.

[57] Shamsad Ahmad. Reinforcement corrosion in concrete structures, its monitoring and service life prediction—a review. *Cement and Concrete Composites*, 25(4):459–471, 2003.

[58] M. Otieno, H. Beushausen, and M. Alexander. Chloride-induced corrosion of steel in cracked concrete - part ii: Corrosion rate prediction models. *Cement and Concrete Research*, 79: 386–394, 2016.

[59] Chong Cao and Moe M S Cheung. Non-uniform rust expansion for chloride-induced pitting corrosion in rc structures. *Construction and Building Materials*, 51:75–81, 2014.

[60] S. J. Williamson and L. A. Clark. Pressure required to cause cover cracking of concrete due to reinforcement corrosion. *Magazine of concrete research*, 52(6):455–467, 2000.

[61] Kim Vu, Mark G Stewart, and John Mullard. Corrosion-induced cracking: experimental data and predictive models. *ACI structural journal*, 102(5), 2005.

[62] D. R. J. Owen, S. Y. Zhao, and E. A. de Souza Neto. *Porous Media: Theory, Experiments and Numerical Applications*, chapter III, pages 437–458. Springer-Verlag Berlin Heidelberg, 2002.

[63] BA Schrefler. Mechanics and thermodynamics of saturated/unsaturated porous materials and quantitative solutions. *Applied Mechanics Reviews*, 55(4):351–388, July 2002.

[64] Bernard D Coleman and Walter Noll. The thermodynamics of elastic materials with heat conduction and viscosity. *Archive for Rational Mechanics and Analysis*, 13(1):167–178, 1963.

[65] William G Gray and S Majid Hassanizadeh. Unsaturated flow theory including interfacial phenomena. *Water Resources Research*, 27(8):1855–1863, 1991.

[66] Wolfgang Ehlers and Chenyi Luo. A phase-field approach embedded in the theory of porous media forthe description of dynamic hydraulic fracturing. *Computer methods in applied mechanics and engineering*, 315:348–368, 2017.

www.ingramcontent.com/pod-product-compliance
Lightning Source LLC
LaVergne TN
LVHW041708190726
843493LV00007B/1991